The David & Charles Manual of

HOME ELECTRICS

Geoffrey Burdett

REVISED BY DAVID HOLLOWAY

The David & Charles Manual of
HOME ELECTRICS

Geoffrey Burdett

REVISED BY DAVID HOLLOWAY

David & Charles

A catalogue record for this book is available from the British Library.

    ISBN 0 7153 0018 0

    First published 1981 (ISBN 0 7153 8112 1)
    Reprinted 1982, 1984
    Second edition 1992 (ISBN 0 7153 0018 0)

© Text Geoffrey Burdett 1981;
    Geoffrey Burdett & David Holloway 1992

© Line drawings David & Charles 1981, 1992

Original illustrations by Steve Pritchard; revisions and additional diagrams by Tom Cross.

The right of Geoffrey Burdett to be
identified as the author of this work has
been asserted in accordance with the
Copyright, Designs and Patents Act 1988.

All rights reserved. No part of this
publication may be reproduced, stored
in a retrieval system, or transmitted,
in any form or by any means, electronic,
mechanical, photocopying, recording or
otherwise, without the prior permission
of David & Charles.

Typeset by XL Publishing Services Nairn
and printed in Great Britain by Butler & Tanner Ltd
for David & Charles
Brunel House Newton Abbot Devon

Contents

	Introduction	7
1	Circuit fuses and MCBs	11
2	Fitting flexible cords	18
3	DIY housewiring	28
4	Light fittings	38
5	Lighting circuit projects	54
6	Lighting controls	80
7	Socket outlets	104
8	Small fixed appliances	122
9	Bells, buzzers and chimes	137
10	Cookers and showers	142
11	Immersion heaters and storage heaters	151
12	Electricity outside the house	160
13	Rewiring a house	170
14	Mainswitches and earthing	178
15	Electricity supply requirements	191
16	Safety in home electrics	196
	Index	203

Introduction

by Geoffrey Burdett

The average intelligent man or woman with a mechanical aptitude and some experience in doing various household jobs from wall-papering a room to laying wall tiles is capable of carrying out electrical work effectively and to the satisfaction of himself or herself and of the relevant authorities. However, in general, for most people any electrical jobs in the house are confined to replacing electric light bulbs, mending fuses and fitting plugs on to the ends of flexible cords. Anything beyond this is apt to be regarded as a mystery and something which should be left to the trained electrician. I would, therefore, like to make it clear from the outset that there is no mystery or mystique about electrical work as applied to the conventional home. As I will illustrate throughout this manual, electrical installation in the home consists largely of a variety of wiring accessories connected to cables fixed to the house structure.

Installing the cables, mainly PVC-sheathed – reasonably flexible and easy to handle – is much simpler than installing the copper pipes and fittings of a plumbing system. Unlike a plumbing system, there is no problem of leakage at the joints of cables and far less cutting away of the structure when putting them in. 'What about leakage of electric current?' the sceptic might ask. Current leakage is certainly a factor in home electrics but as explained in the relevant sections of the manual it is as much a question of the careful use of electrical appliances as of the installing of fixed wiring. It may surprise readers of little or no experience with electrical wiring that more than ninety per cent of home electrical installation work is in fact *non*-electrical. The non-electrical chores include moving furniture, lifting floor coverings, raising floorboards and refixing them, drilling joists and walls, chopping out plaster and masonry and making good, running and fixing cables, fixing wiring accessory boxes together with the mechanical fixing of switches, light fittings, socket outlets and other wiring accessories. All these tasks are hardly different from the ordinary jobs carried out by the DIY man or woman and need fewer skills than, say, woodworking or plumbing. But all these non-electrical tasks, when carried out by the tradesman electrician, cost as much in labour as does the most intricate electrical job.

The actual electrical work in home wiring consists mainly of preparing the ends of cables and flexible cords and connecting the wires to the terminals of the wiring accessories, electrical appliances and apparatus, which in some instances includes connections at, or the complete installation of, a consumer unit.

For the inexperienced and even for the experienced I should point out that especial care and some skill is required in stripping off the appropriate amount of sheathing and insulation from cables and flexible cords. It is necessary to ensure that the insulation of the conductors is not damaged by the knife or stripping tool and that there are no exposed portions of uninsulated wires within an accessory or appliance likely to produce short circuits or other faults. Care must also be exercised when dealing with terminals to ensure that conductors make effective contact, with no loose wires leading to excessive temperature rise and possibly fire.

Modern wiring accessories in the main are well designed and of good quality, with easy wiring facilities so the do-it-yourselfer will have little if any difficulty in connecting the wires to the terminals and fixing the accessories to their appropriate mounting boxes. Nevertheless it is important that the terminals of each accessory will accept the number and sizes of cables required without difficulty. Careful planning of the wiring, especially of extensions to circuits, is necessary as recommended in the manual. Connecting each wire to its appropriate terminal is also very important. It is in junction boxes in particular where attempts are made to include more cables than the box is designed to accommodate with the result that the terminals will not accept all the wires and the box covers cannot be replaced. The principal reason for overcrowding is that the junction boxes are too small in the first place for the size of the project.

Much the same applies to ceiling roses in which three sheathed cables is the norm but one and only one additional 2-core and earth sheathed cable for an extension light can be accommodated Attempts to include more produce problems.

Socket outlets and their boxes also have limited accommodation for cables, one spur cable in addition to the two ring cables being the desired maximum; but by careful planning the use of double socket outlets in deeper boxes makes wiring easier and the connection of wires more effective.

Much of the work in the manual deals with extensions to the various circuits to provide additional lights and socket outlets; also various modifications to update an installation. These are the jobs on which most householders require and seek information.

Most of the tasks are comparatively simple but before extending any circuit and even before planning the cable route, check that any additional point or outlets will not exceed the regulation limits as laid down in the manual – otherwise a circuit could be overloaded.

From my own experience I am aware that the requirements of a high proportion of householders are for shower unit installations, electric cookers and an electricity supply to a detached garage, a greenhouse and/or shed. These extensions are somewhat different in that they require entirely *new* circuits which in turn mean additional fuseways in the consumer unit. Since most consumer units have no spare fuseway or if one is present it has probably long been utilised for a new circuit, consideration has to be given to the provision of at least one extra fuseway. As explained in the manual there are a number of ways of doing this but from the outset it must be realised that there is little point in buying the shower or cooker without first providing the facilities for the necessary circuits.

With regard to supplying electricity to an outhouse such as a garage or greenhouse this must not be done via a flexible cord plugged into the kitchen socket outlet or other convenient 'temporary' point. It must be permanent wiring properly installed, using the correct cable for outside work.

Running an electricity supply to the loft is another job that most householders intend to carry out but rarely do so; this is dealt with at length in the manual. Where a house is being extended or a conservatory or sunroom is erected and fixed to the rear of the dwelling the electrical work is merely an extension of the existing power and lighting circuit.

Buying materials

Although most brands of wiring accessories are of good quality it is always advisable to choose those of leading makes and from makers specialising in certain items. These are sold at prices usually no higher and sometimes lower than those of inferior quality. By choosing the best you get not only a high quality product but the accessories are usually easier to install and the terminals are more generous in cable capacity, so ensuring first-class electrical connections.

Similar differences do not apply to cables. There are fewer makes and *all* house wiring cables are made to the relevant British Standards and are of good quality. Cables are rarely imported, but if they are available the do-it-yourselfer should avoid them. Imported wiring accessories should also be eschewed. Our 240-volt electricity supply is the highest in the world applying to domestic interiors, and British-made accessories and cables are designed to cover that voltage with a generous margin.

Many retail suppliers sell cables and wiring accessories at discount prices and provided that these items are by recognised manufacturers, shopping around is well worth while. The best prices will often be found in the do-it-yourself 'superstores' where you will be able to buy in bulk – cable in 50m or 100m rolls and 10 socket outlets at a time, for example. But you will not always be able to buy the more specialised accessories, such as whole-circuit RCDs (residual current devices), and for these the best bet is an electrical wholesaler who, these days, will normally be quite happy to serve the do-it-yourselfer as well as the tradesman. Many shops specialising in electrical domestic appliances tend to stock only a few accessories and consequently charge maximum prices.

Workmanship

Good workmanship is essential in all electrical work. As a rule the DIY person takes a pride in what he is doing, takes his time and usually produces work of a high standard; there is no reason why such standards should not apply equally to his electrical pursuits.

The DIY 'electrician', however, is frequently blamed for fires attributed to faulty wiring, most often without justification – the real cause being the presence of old wiring which should have been renewed. Old wiring having rubber insulation is highly combustible, whereas PVC does not support combustion. It is therefore unlikely that recent wiring would be the cause of a fire, no matter who installed it.

There *are* instances of bad wiring installed by DIY handymen who are not conversant with the requirements of a good installation. The main purpose of this manual is to show how, and what, constitutes recognised good practice and how to do the work correctly. However, much bad wiring is carried out by less reputable 'electrical contractors', often unscrupulous individuals known in the trade as 'cowboys', who use door-step trading methods and scare householders into having their homes rewired on some pretext or other. I know these 'moonlighters' not to be trained electricians but in many instances are lorry drivers, dustmen, postmen and even insurance collectors out to make a 'quick buck' at the expense of the householder.

There are some trained electricians who, unknown to their employers, carry out wiring jobs in their spare time. These operators normally use higher quality materials and do a better job, though their main aim is to complete the work in the shortest possible time and accept only cash in payment.

The DIY person following the instructions in this manual should be able to produce a first-class job. A sensible precaution, however, is to have wiring installations regularly *tested* by a professional electrician, who will have specialised equipment.

Permission to install wiring

There are no laws in England or Wales preventing or restricting a person from carrying out electrical work in his own premises, as there are in some countries. As has already been explained, cables and accessories for house wiring are readily available from reputable electrical accessory suppliers so it is just a question of choosing the right materials and knowing how to use them.

Attempts are continually being made by electrical contractors and other interested parties to obtain legislation to restrict the installation of house wiring to registered contractors and electricians, and so outlaw DIY electrical work. This is unlikely to be introduced. The only possible way of imposing this restriction would be for the trade to stop retailing cables and accessories to the general public, but this is also unlikely.

Attempts have also been made in the contracting industry to persuade electricity companies to refuse to connect electrical wiring to the mains unless installed by a registered electrical contractor presenting a certificate of inspection and test.

Electricity supply undertakings are required by law to connect an electrical installation to the mains unless it is dangerous and is contrary to the requirements of the Electricity Supply Regulations. They will be looking for compliance with the IEE (Institute of Electrical Engineers) *Regulations for Electrical Installations* – or the 'wiring regulations'. In Scotland, these are part of the building regulations, so have the force of law.

Domestic electrical appliances

Except for renewing a flexible cord, replacing the belt of a vacuum cleaner, the carbon brushes of an electric motor, the element of an electric kettle or an electric fire, and other small jobs, repairs to domestic electrical appliances should not be attempted without the aid of the maker's servicing manual for the specific model. These manuals are not however usually available other than to authorised mechanics who have completed a training course on each model. Also, special tools are needed and many components are press fitted so that the job cannot be done in the home. Spare parts for replacement too are not generally available. With similar appliances the tendency is to design them as unrepairable 'throwaway' products, because of the high cost of labour for servicing. With the larger appliances a service call is charged at £30 or more for the first fifteen minutes, which is especially costly when the 'fault' is found to be only a blown plug fuse. One solution is to negotiate a servicing contract once the period of guarantee has expired.

Because of these factors the manual does not attempt to cover appliance repairs nor does it encourage the householder to tamper with his appliances. An attempt by the householder to rectify a small fault, without consulting the appropriate servicing manual or using the recommended special tools, could result in a major repair at greatly increased cost.

Geoffrey Burdett 1981

Editor's footnote

Since Geoffrey wrote this introduction for the 1981 edition, not a lot is different for the home electrician as far as carrying out the work is concerned, but there have been other changes.

There have been two new editions of the IEE regulations – the current edition is the 15th, but this will be superseded by the 16th Edition on 1 January 1993 and changes in the 16th Edition have been taken into account in the revision of this manual. It would certainly be worthwhile for the keen handyman or woman to obtain their own copy of the wiring regulations, though some of the guides available are easier to read.

The biggest change has been in the accessibility to the tools, materials and equipment required – all because of the mushrooming growth of the do-it-yourself superstores. Many of these sell a good selection of basic electrical supplies and will often have in-store information on various electrical jobs.

Some manufacturers have designed electrical tools, such as wire strippers, specifically with the do-it-yourselfer in mind and many wiring jobs are made easier (and safer) with the introduction of cordless electric drills which enables holes to be made in walls and in flooring joists without the need for an electrical supply. There is also a good range of inexpensive test equipment available – though not the more specialised equipment used by professional electricians for testing things like earth loop impedance.

Safety has become more important and you can now buy plug-in portable RCDs and RCD-protected socket outlets (see page 201 for more details) as well as 'split-load' consumer units with an RCD protecting half or more of the circuits.

David Holloway 1992

IMPORTANT NOTE

Before carrying out any electrical work, turn off the power. It is best to remove the circuit fuse in case someone else turns the power on again at the mainswitch inadvertently, or where it is desirable and safe to restore the power to the remaining circuits (for freezers, fridges, etc). Before working on portable appliances, pull out the plug.

1: Circuit fuses and MCBS

Each electrical circuit in the home has a fuse or MCB (miniature circuit breaker) to prevent the circuit cable overheating, as a result of a fault or serious overload, and causing a fire. Plugs are similarly fitted with fuses to protect the flex.

Most existing installations have circuit fuses, located in the consumer unit or fuseboard, though MCBs, which are more effective and more convenient, are becoming increasingly popular.

Circuit fuses

There are two basic types of circuit fuses: (i) rewirable fuses and (ii) cartridge fuses.

The rewirable fuse is fitted into the majority of consumer units (see Fig 1) and fuseboards but the cartridge fuse is the more superior and requires less current to blow it than does a rewirable fuse of the same current rating. The current rating of the fuse determines the circuit current rating and the minimum size of cable for the circuit. The current rating of a fuse is the maximum amount of current it will carry continuously without undue rise in temperature. The amount of current required to blow a fuse is much higher than the rating. For a rewirable fuse a current twice the rating of the fuse is needed and the fuse is said to have a fusing factor of 2. A cartridge fuse blows when the current is $1\frac{1}{2}$ times the rating and therefore has a fusing factor of 1.5. For example a 30A rewirable fuse protecting a ring circuit requires 60 amps to blow it whereas a 30A cartridge fuse needs only 45 amps, thereby placing less strain on the circuit wiring.

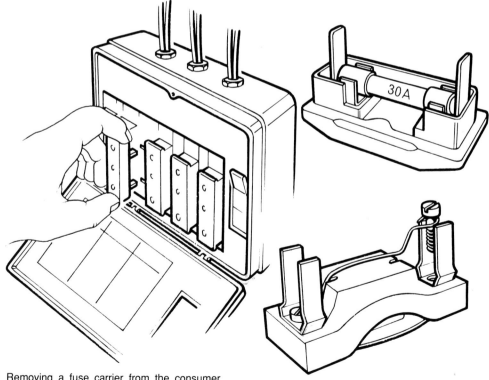

Fig 1 Removing a fuse carrier from the consumer unit. *Top right*: A cartridge fuse in its fuse carrier. *Bottom right*: A rewirable fuse widely in use but now obsolescent.

12 Circuit fuses and MCBS

Colour coding

Circuit fuses are colour coded according to their current rating: 5A white, 15A blue, 20A yellow, 30A red and 45A green.

Rewirable fuses

A rewirable fuse consists of a short length of tinned copper wire connected to terminals in a fuse carrier. At each end of the carrier is a single or a double knife-blade contact which is inserted into corresponding spring contacts in the fuseway of the consumer unit or fuseboard containing the fixed portion of each fuse unit.

The fuse carrier is colour coded as is also the base of each unit; the units are so designed that it is not possible to insert a fuse carrier of a given current rating into the fuseway of one having a lower current rating, with one exception: a 15A fuse has the same dimensions as the 20A. These are interchangeable but the small difference in current rating makes this of little consequence.

Advantages of rewirable fuses

The advantages of rewirable fuses are that they are cheap to buy and cost hardly anything to 'mend'; fuse wire is usually readily available.

Disadvantages of rewirable fuses

The rewirable fuse requires more current to blow it. Sustained overload erodes the tin on the fuse wire, reducing its effective current capacity and causing the fuse to run hot and damage the insulation of the circuit cable. It is subject to abuse by fitting larger fuse wire having a higher current rating – even nails have been used as substitutes – with the result that the circuit is severely overloaded.

Cartridge fuses

Cartridge-type fuses are similar to rewirable fuses except that the fuse carrier contains a cartridge instead of a fuse-wire element.

A cartridge fuse is a ceramic tube containing a silver fuse element packed in quartz powder (sand) which is connected to a silvered contact at each end. The contacts make a metal-to-metal electrical contact with those of the fuse carrier.

Advantages of cartridge fuses

This type of fuse does not deteriorate in use. It requires less current to blow, and when it does blow it does so with greater speed than a rewirable fuse. Different current ratings have different sizes of cartridge, making it impossible to uprate in use.

Disadvantages of cartridge fuses

They are comparatively expensive to buy. Spare fuses of *each* current rating have to be available. They cannot be rewired. It is not possible to tell by visual inspection when a fuse has blown. Absence of spare cartridges tempts householders to insert substitutes or attempt to rewire a cartridge with sometimes disastrous results.

Mending fuses

For mending fuses the materials and tools required are: a card of fuse wire giving different current ratings, or at least two cartridge fuses of each current rating in the consumer unit; a small screwdriver, a pair of pliers, sidecutters and a torch.

When a fuse blows take the torch and turn off the mainswitch. Remove the fuse cover and check the circuit list to see which fuse has blown. If no circuit list is present, it will be necessary to remove in turn each fuse of a particular current rating where there are more than two of that rating. For example, if the blown fuse is in a lighting circuit, each fuse colour-coded white may have to be checked.

If it is a rewirable fuse, check the fuse wire in the carrier. If it is a cartridge fuse insert the suspect fuse into the fuseway of a circuit known to be working. Otherwise fit a known sound fuse and test the suspect fuse after the power is restored.

Mending a rewirable fuse

Take the fuse carrier containing the blown fuse (see Fig 2). Loosen the two screws and take out the pieces of old fusewire. Also clean off any blobs of melted copper and tin.

Select a piece of fuse wire of the correct current rating and thread the end into the tube

Circuit fuses and MCBS 13

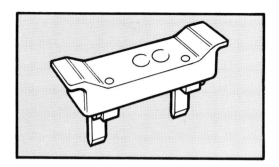

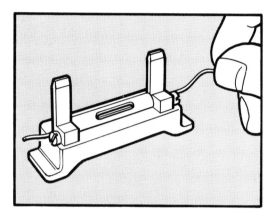

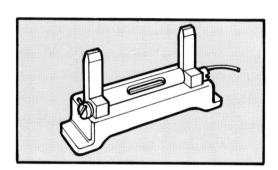

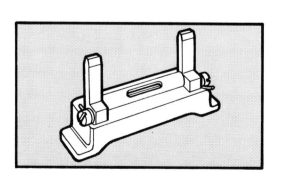

Fig 2 Stages in 'mending' a rewirable fuse.

of the fuse carrier. Connect it to the terminal. If a clamp terminal, place the end of the wire under the washer clockwise and tighten the screw. If a screw-hole type, double back the end of the fuse wire, insert it into the terminal hole and tighten the screw.

Cut the wire to length and connect the other end to the other terminal making sure you do not stretch the wire; it is preferable to leave a little slack.

Replace the carrier in the consumer unit and turn on the mainswitch. If the fuse blows again immediately there is a serious fault in the circuit which needs to be located and rectified before renewing that fuse.

Mending a cartridge fuse

Take out the suspect fuse and test it with a special fuse & bulb tester or a simple battery-operated continuity tester containing its own tiny light bulb. Hold the tester's probes against each end of the fuse; if the fuse has blown, the bulb will not light. Having ascertained which fuse has blown, fit in a new cartridge of the same current rating. Replace the carrier and turn on the mainswitch. If the fuse blows again immediately there is a serious fault in the circuit.

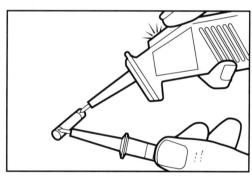

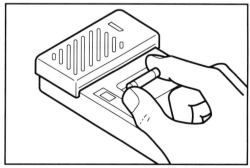

Fig 3 *Top*: Testing a cartridge fuse using a continuity tester. *Bottom*: Using a fuse & bulb tester.

Photo 1a 'Split-load' consumer unit fitted with MCBs with some circuits protected by an RCD (*MK*).

Miniature circuit breakers

The miniature circuit breaker (MCB) is a single-pole switch which cuts off the supply automatically when excess current flows in the circuit and through the tripping device of the MCB.

MCBs are fitted into their own consumer unit and have similar ratings to fuses: 6A, 16A, 20A, 32A and 45A. MCBs and fuses are not usually interchangeable, though at least one manufacturer (*Wylex*) makes plug-in MCBs to replace existing fuses.

MCBs operate with lower current fault than do fuses and with greater speed. Only $1^1/_4$ times the current rating of an MCB is needed to trip it, this being equivalent to a fusing factor of 1.25. For example a 20A MCB will trip at 25A compared with 30A for a 20A cartridge fuse and 40A for a rewirable fuse.

The advantages of an MCB are considerable since only a comparatively low excess current needs to flow in the circuit to operate it and the reaction is almost immediate. Also, individual circuits can be switched off if desired. For example, when leaving the house unoccupied for a period most of the circuits can be switched off with the exception of those supplying the freezer and, in the winter, the central heating.

The disadvantage of the MCB is its high initial cost, but this is recouped in time and money saved by not having to replace fuses.

Circuit fuses and MCBs 15

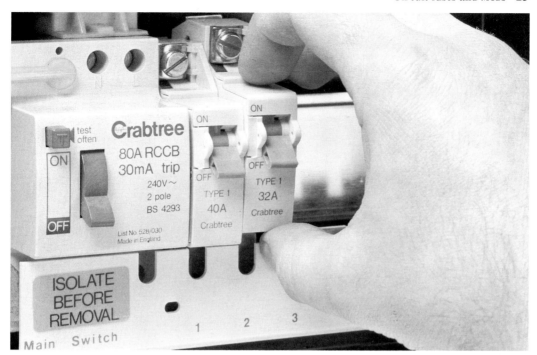

Photo 1b Clipping an MCB on to the 'busbar' of a consumer unit (*Crabtree*).

Photo 2 A consumer unit fitted with MCBs and protected by a whole-house RCD (*Wylex*).

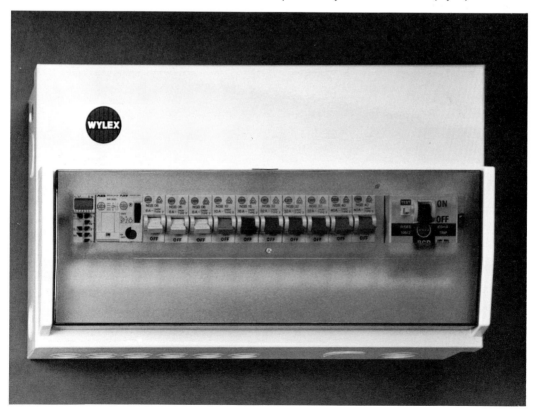

16   Circuit fuses and MCBS

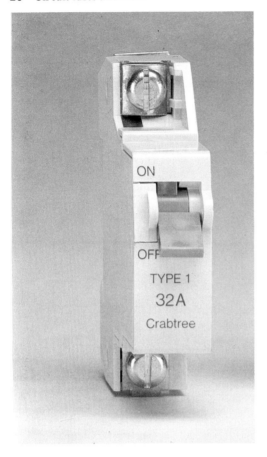

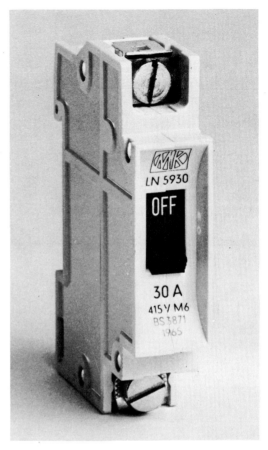

Photo 3a Switchtype MCB (*Crabtree*); **Photo 3b** Switchtype MCB (*MK*); **Photo 3c** Combined MCB and RCD for consumer units (*Wylex*).

Action to be taken when an MCB operates

When an MCB operates the lights or power on the circuit fail in the same way as when a fuse blows. When the supply fails in a circuit, inspect the consumer unit where it will be seen that the switch on the MCB unit is in the OFF position.

First try to ascertain what caused the failure – overload or a fault. If located, first have the fault rectified or the appliances responsible for an overload switched OFF.

Turn on the switch of the MCB. If it stays 'on' all is well. If you cannot close (switch on) the circuit breaker the fault remains and must be rectified before the current can be restored to the circuit.

Why fuses blow or MCBs operate

We have seen that a circuit fuse has a given current rating from which can be derived the current required to blow the fuse or MCB.

A fuse blows mainly for one of two

reasons: (i) short circuit, where a live wire is in contact with the neutral with no appliance or other resistance to limit the current, the only limitation being the amount required to blow the smallest fuse in the circuit; and (ii) where current is leaking to earth, the amount of which is equal to that required to blow the smallest fuse. Leakages of this magnitude occur where the live wire is in contact with the earth wire or with earthed metalwork. Where an RCD is fitted (see below), this will trip and so cut the current before the fuse blows.

A fault in an appliance will normally blow the plug fuse, not the circuit fuse.

A lighting-circuit fuse is likely to blow for any number of reasons. Even a failed bulb may blow the lighting-circuit fuse or MCB.

Residual current devices (RCDs)

An RCD is *not* the same as a fuse or MCB, though many people confuse the two – partly because many electricians still call an RCD by its old name of earth-leakage circuit-breaker or residual-current circuit-breaker.

What an RCD does is to monitor the currents flowing in the live and neutral wires and to 'trip', switching off the current, if it senses any difference, which could be because of a leakage of current to earth. It will trip at very small currents and will operate very quickly, preventing you getting an electric shock if you are in contact at any time with a live electric wire, directly or indirectly.

RCDs can be fitted in the plug, in an adaptor fitted between plug and socket, in the socket, to a single circuit, in the consumer unit to protect several circuits or before the consumer unit, protecting the whole house. For more details, see Chapter 14 and page 201.

Combined MCB/RCD units are available for fitting to consumer units to provide RCD protection for a single circuit.

Fuses and colour code

Table 1: **Circuit fuses**

Current rating (Amps)	Colour	Circuit function
5	white	Lighting
15	blue	Immersion heater; single 13A or 15A socket outlet
20	yellow	Storage heater; multi-outlet radial circuit (13A outlets)
30	red	Cooker; ring circuit; radial circuit (13A outlets); electric showers up to 7.2 kW
45	green	Cooker (very large family-size); electric showers up to 10.8 kW

Table 2: **Plug fuses**

Current rating (Amps)	Colour
2	black
3	red
5	black
10	black
13	brown

Plug fuses

The cartridge fuses in plugs protect the flex leading to the lamp or electrical appliance and also the appliance itself. Although there are several ratings theoretically available (see Table 2), most plugs are sold with fuses rated at 3A (for use with appliances up to 720W) or 13A (for use with appliances up to 3kW).

Fused connection units, used for the permanent wiring of small fixed appliances, are fitted with the same fuses as plugs.

2: Fitting flexible cords

When fitting a new flex (flexible cord), it is important to make sure that it is the correct type and size. In most instances the correct replacement flex will be the same as the old flex but there are instances where the old flex which is being replaced is the wrong type or the wrong size.

The size (that is, the cross-sectional area of the conductors) of flexible cord ranges from $0.5mm^2$ the smallest with a current rating of 3A to $1.5mm^2$ with a current rating of 16A, though larger sizes (up to 32A) are available.

Types of flexible cord

Of the numerous types of flexible cord available there are five which meet most if not all requirements in the home. These are: (i) circular PVC-sheathed flex; (ii) heat resisting rubber-sheathed flex; (iii) braided circular flex; (iv) unkinkable flex, and; (v) parallel twin flex.

Circular PVC-sheathed flex

Circular PVC-sheathed flexible cord is made in 2-core and 3-core versions, having either two or three PVC-insulated conductors. This type is the most widely used, being hard-wearing and able to withstand a considerable amount of rough handling without damage. It is fitted to many types of electrical appliance, including power tools, mowers and hedge-trimmers and is available in numerous colours of sheathing, white, black and also grey being mostly used for appliances; orange and other bright colours for garden tools, such as lawnmowers and hedgetrimmers, and extension leads.

For pendant light fittings, a heat-resisting PVC sheathed flex is available; you can also get 'curly' PVC-sheathed flex, often already wired up as an extension lead or a kettle connector, and flat twin 2-core PVC-sheathed flex.

Heat-resisting rubber-insulated flex

Circular heat-resisting flex has rubber insulation on the cores and is covered with a tough rubber sheath. It is used where high temperatures could be a problem, such as the flex leading to immersion heaters or storage heaters.

Braided circular flex

Braided circular 3-core flexible cord contains three rubber-insulated conductors and textile fillers to provide the circular cross-section, with an over-all two-colour braiding of cotton. This type of flex is fitted to portable electric heaters and appliances where there is little if any risk of mechanical damage.

Unkinkable flexible cords

Unkinkable flexible cord, mostly 3-core, has rubber-insulated conductors and textile fillers contained in a thin rubber circular sheathing with an overall two-colour braiding moulded on to the rubber. This type of flex is fitted to electric irons, kettles, coffee percolators and similar appliances which could be adversely affected by high temperatures at the terminals if PVC insulation were used. It has largely given way to circular sheathed flex of which there is a choice of rubber or PVC insulation.

Parallel twin flexible cord

Parallel twin flexible cord is used for the wiring of fixed light fittings where the flex is wired into the fitting and does not support a lampholder and shade, as in the case of a plain pendant light fitting. This type of flex may also used for table lamps and smaller electric appliances such as electric clocks, but has now largely been superseded by *flat twin* PVC-sheathed flex.

Fitting flexible cords to appliances

When a flexible cord is damaged or shows signs of wear replace it. Do not attempt to repair flex which is damaged or has worn sheathing.

To disconnect and remove the old flex, access to the terminal block of the appliance is necessary. When the terminal block is located and the cover removed, disconnect the old flex wires noting the terminals for the three colours. The 'L' terminal may have a spot of red, the 'N' a spot of black and the 'E' a spot of green paint. These were the former code colours and most likely the core colours of the old flex. The new flex has the current code colours: brown for *live*, blue for *neutral* and green/yellow for *earth*.

Thread the end of the flex through the cord entry hole which, if a metal casing, will contain a grommet. Prepare the end of the flex and connect the brown wire to the 'L' terminal, the blue wire to the 'N' terminal, and green/yellow wire to the 'E' terminal. A double-insulated appliance indicated by a double hollow square will have no 'E' terminal and its flex will be 2-core.

Anchor the sheath of the flex in the cord clamp near the terminal block. Replace the terminal block cover and any other items removed to facilitate the flex renewal. Fit a plug to the end of the flex (see Fig 7).

The procedure for fitting a flex varies considerably with the make and model of the appliance. It is therefore necessary to examine each appliance carefully and where available obtain a copy of the makers' instructions.

With some appliances, especially small ones, it is sometimes impossible to gain access to the terminals without wholly or partly dismantling the appliance itself. In this case do not attempt to replace the flex yourself. Instead, take it to an appliance repair shop.

Electric kettles

Electric kettles have a removable flex adaptor which is 'plugged' into the body of the kettle containing the element pins (see Fig 4). On older kettles the adaptor is rewirable; on modern kettles it is moulded on so the whole flex has to be replaced.

Electric irons

Most models and makes of electric irons have the flex connected to a terminal block housed in the iron's body. Fig 5 shows the terminal configuration and the flex connections of an automatic electric iron which is fairly representative.

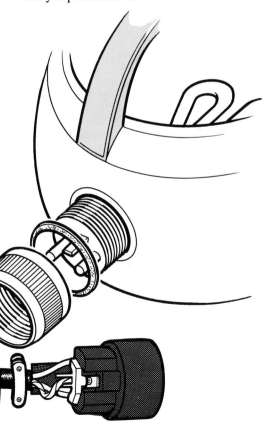

Fig 4 *Top*: A new element has been inserted into the kettle and the external fibre washer is placed on the flange followed by the shroud which is screwed tight to prevent leaking. *Bottom*: The flex wires have been connected to the terminals of the kettle connector and the cord grip secured to the sheathing.

20 Fitting flexible cords

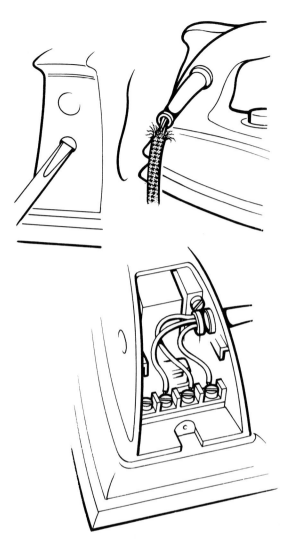

Fig 5 The terminal cover at the rear of the iron is removed, the old flex disconnected and the new 3-core flex connected as shown.

Extending flexible cords

When a longer flexible cord is required for an appliance an entirely new length should be fitted. Flexible cords should be as short as possible, normally 2m for electric fires and other such appliances. A length of 3m is acceptable for an electric iron, whereas a much longer section of flex is fitted to appliances such as vacuum cleaners. Garden tools, such as lawnmowers and hedge-trimmers, have a short flex attached to the machine, which connects to a much longer extension lead. Power tools generally should have fairly short flexible cords; where a longer

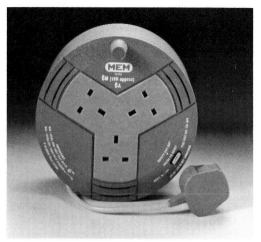

Photo 4 6A extension reel with three socket outlets, safety cut-out and moulded-on safety plug (*MEM*).

flex is occasionally required a flexible cord extension should be used, preferably contained on a drum fitted with a socket outlet and neon indicator.

When buying a flexible cord extension reel choose one fitted with 3-core flex, so that when it is used with an appliance or power tool requiring earthing the earthing will remain intact. Also ensure the flex is of a current rating suitable for the appliance with the highest loading to be run off the extension reel: ie, 1.5mm^2 or at least 1.25mm for 3kW equipment. Many extension leads, however, have ratings of only 5A, suitable for equipment up to 1200W.

Do not leave any flex coiled on the reel when using the extension. The current rating of the flex applies to its use in free air. If any flex is left coiled, heat is generated and likely to start a fire.

Flex connectors

Where a flex has to be extended and an extension reel not employed, there are two ways of making the join in the flex: (i) by means of a connector (see Fig 6) containing three terminals to which the flex is connected permanently, as in a junction box (waterproof versions are available); and (ii) using a 2-piece connector, one piece being a socket and connected to the mains lead, the other piece having pins which connect with the socket section, which must be joined to the appliance flex and *not* the mains flex.

Fitting flexible cords 21

Photo 5a 4-way 13A extension socket with neon indicator (*Duraplug*).

Photo 5b 4-way 13A extension sockets. *Left*: With neon indicator and fuse. *Right*: With neon indicator, fuse and switch (*Duraplug*).

Photo 5c Flex connectors plus spare 2-pin and 3-pin plugs and sockets (*Duraplug*).

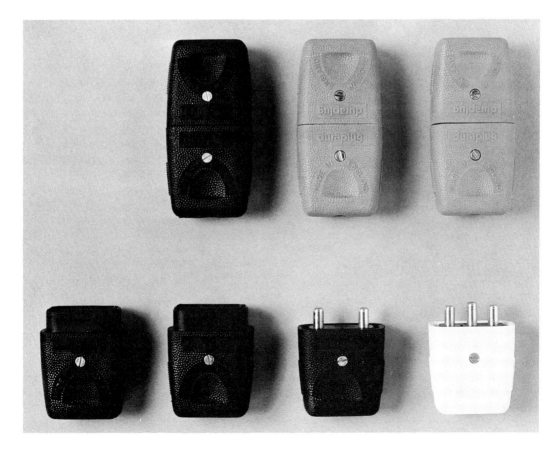

22 Fitting flexible cords

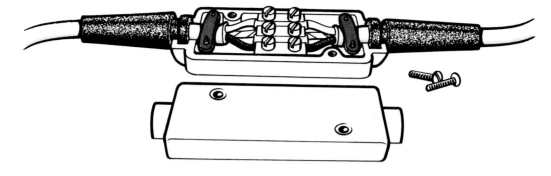

Fig 6 A 3-terminal connector containing cord clamps and grommets and a cover secured by screws is the safest and usually the best method of joining two flexible cords.

Table 3: **Flexible cord sizes and applications**

Conductor size (mm^2)	Current rating (amps)	Application
0.5	3	Lighting fittings
0.75	6	Lighting fittings and small appliances
1.0	10	Appliances up to 2000W
1.25	13	Appliances up to 3000W
1.5	16	Appliances up to 3840W
2.5	25	Appliances up to 6000W
4.0	32	Appliances up to 7680W

Photo 6a Waterproof flex connector for use outside (*PowerBreaker*); **Photo 6b** 'Strip and Fit' tool designed to make wiring plugs easier (*PowerBreaker*).

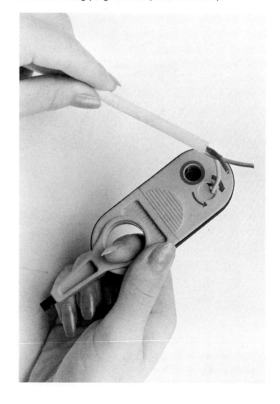

The sizes of flexible cord in the above table are used for the various appliances as indicated, but in practice makers tend to fit 1.25mm^2 to most appliances of up to 3000W for use from a 13A socket outlet.

Wiring plugs

The function of a plug is to connect a portable electrical appliance or portable lamp to the fixed circuit wiring. The plug itself is connected to the end of the flexible cord fitted to the appliance or light. Plugs are designed to accept flexible cords of various types and sizes but *not* ordinary house wiring cables.

Fitting flexible cords 23

Types of plugs

There are two principal types of domestic electric plug: (i) flat-pin fused plugs and (ii) round-pin non-fused plugs.

Flat-pin fused plugs

The standard fused plug used in Britain is the 13A 3-pin plug having rectangular (flat) pins which fits the 13A socket outlet of the domestic ring circuit and other multi-outlet power circuits. The plug and its socket outlet are made to British Standard BS 1363, which number is stamped on each plug and socket outlet. The plug must additionally be marked with a British Standards Kitemark or the symbol of the approving testing authority, such as ASTA.

The 13A fused plug contains a replaceable cartridge fuse secured in two metal spring clips which are the electrical contacts between the fuse-end caps and the plug. The fuses bear a stamp, BS 1362, and are made in a number of current ratings, each of which is colour coded.

Only two current ratings are a regulation requirement for domestic plugs. These are 3A (red) which is used for lamps and appliances having loadings up to 720W; and 13A (brown) for appliances having a loading of from 720W up to and including 3000W. Current ratings of the other fuses are 2A, 5A and 10A all colour coded black. These fuses are used for special purposes but are rarely fitted into plugs used in the home.

All new 13A plugs now have partially-sleeved pins to prevent the possibility of electric shock when the plug is halfway into or out of a socket outlet. In addition to the normal plastic plug (available in colours as well as white), you can get heavy-duty resilient versions for use where the plug might get a lot of knocks – on power tools, for example.

Round-pin non-fused plugs

Round-pin non-fused 3-pin plugs were widely used in house wiring before the advent of the 13A flat-pin plug. Three versions were used – 15A, 5A and 2A – each on its own separate radial circuit.

The two larger sizes have now virtually disappeared, though 2A plugs are making a

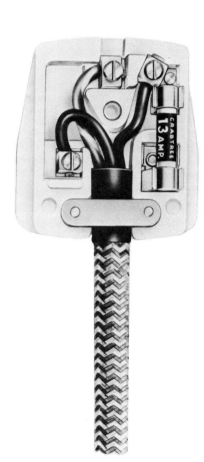

Photo 7a A 13A plug with insulated sleeves (*Crabtree*); **Photo 7b** Inside view showing flex connections and cord clamp (Crabtree).

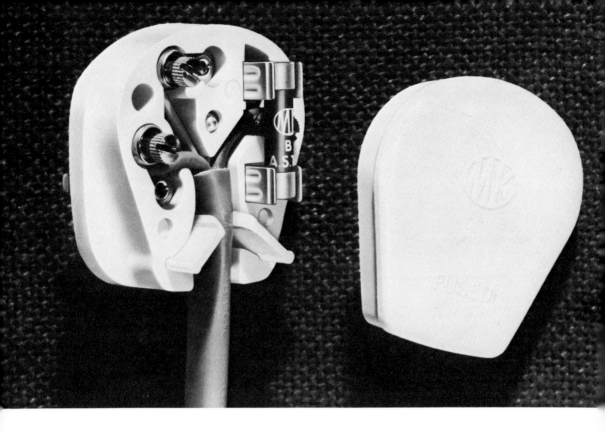

Photo 7c Inside view of 13A 'Safetyplug' with push-in cord grip and pillar terminals (*MK*); **Photo 7d** 'Easywire' 13A plug with colour-coded terminals, push-in cord grip and hinged lid – the fuse can be accessed externally (*PowerBreaker*).

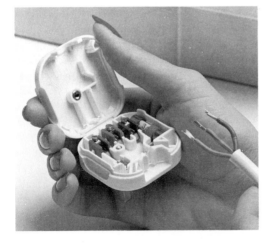

comeback for use in lighting circuits, so that several table lamps can be brought on simultaneously from a light switch by the door (see page 175).

The other type of round-pin non-fused plug you will come across is the two-pin version on electric shavers, which plugs into a special adaptor or a shaver socket. These are now all moulded on to the flex and cannot be replaced.

Wiring a 13A fused plug

First loosen the screw on the face of the plug securing the cover.
(i) Lift off the cover to give access to the terminals and the fuse (see Fig 7). With the cord outlet at the bottom and the fuse on the right, the terminal next to the fuse is the live, marked 'L', the terminal on the left is the neutral, marked 'N', and the top terminal is the earth, marked 'E'. Take out the cartridge fuse.
(ii) Place the flex over the cord grip and measure the amount of outer sheathing to be removed, which for most plugs is 38mm.
(iii) Strip off the required length of outer sheathing by bending the flex over your

Fitting flexible cords 25

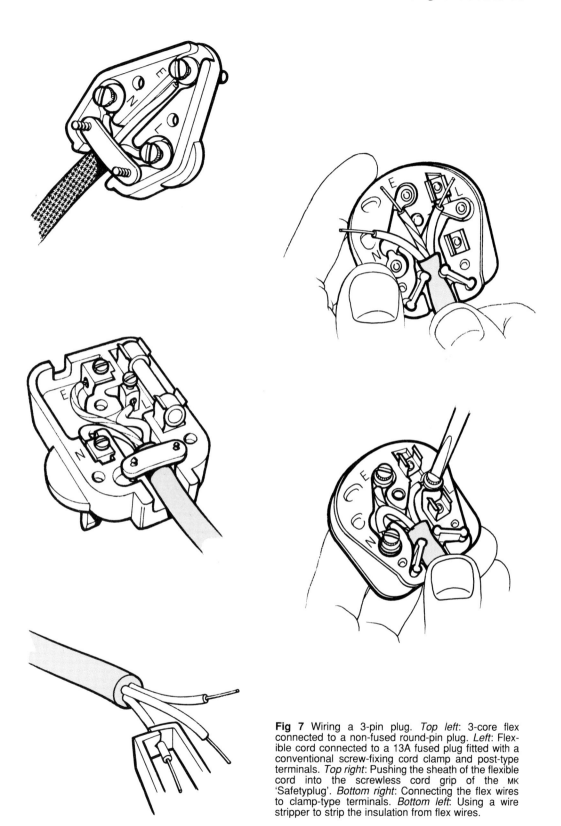

Fig 7 Wiring a 3-pin plug. *Top left*: 3-core flex connected to a non-fused round-pin plug. *Left*: Flexible cord connected to a 13A fused plug fitted with a conventional screw-fixing cord clamp and post-type terminals. *Top right*: Pushing the sheath of the flexible cord into the screwless cord grip of the MK 'Safetyplug'. *Bottom right*: Connecting the flex wires to clamp-type terminals. *Bottom left*: Using a wire stripper to strip the insulation from flex wires.

26 Fitting flexible cords

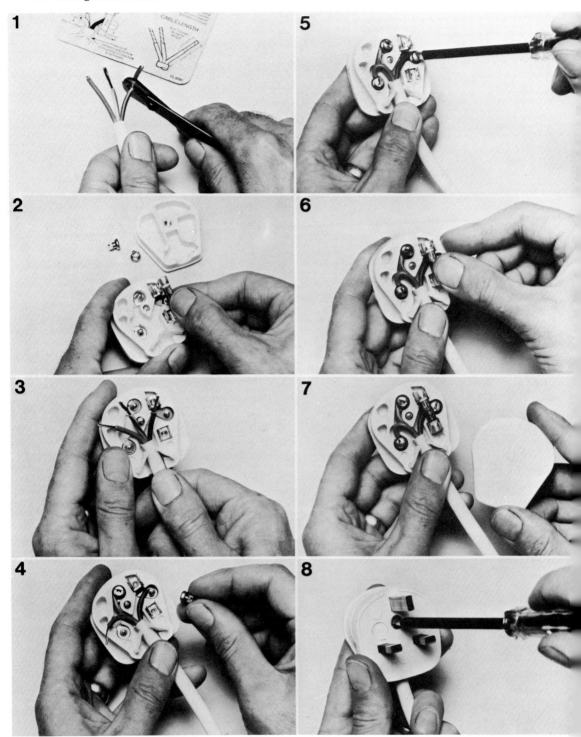

Photo 8 Eight stages in fitting flex to a 13A plug: **1** Stripping the equal length wires; **2** Removing the fuse; **3** Pushing the flex into the cord grip; **4** Attaching the terminal heads by hand; **5** Tightening with a screwdriver; **6** Replacing the fuse; **7** Replacing the plug cover; **8** Tightening the securing screw (*MK*).

finger and making a circular cut all round the flex, into but not quite through the sheathing. Bend the flex both ways to break the last little bit and pull off the sheathing. Now lay the flex on the plug, such that the sheathing is well inside the cord grip and, if necessary, cut the individual wires to reach around 12mm beyond each terminal (some plugs are designed to accept equal lengths of wire in the three terminals). Finally, strip about 12mm of insulation from the end of each of the three wires – note that the soldered ends on wires in flexes are there for the manufacturer to test the appliance in the factory and should be cut off before you fit a plug.

Connect the green/yellow wire to the 'E' terminal, connect the brown wire to the 'L' terminal, and finally connect the blue wire to the 'N' terminal.

There are two types of terminal: (a) the wire hole post terminal and (b) the clamp terminal.

If a post terminal, loosen the terminal screw, insert the bared end of the appropriate wire then tighten the terminal screw. (For smaller flexes the bared ends of the wires should first be doubled to increase the contact area and the mechanical strength of the connection.)

If a clamp terminal, loosen the screw and place the bared end of the wire clockwise under the washer and tighten the screw. In the MK 'Safetyplug' the terminal head with its captive washer is removable, the wire is placed round the fixed terminal screw, the milled and slotted screw head is then replaced and tightened by means of a screwdriver.

(iv) Position the end of the sheathing under the clamp with the maximum length of sheathing projecting into the plug body and tighten the clamp screws so they grip the flex. The MK 'Safetyplug' has a screwless cord grip in the form of an inverted V and the sheathing of the flex is pressed into the grip. With this type, the greater the strain on the flex the tighter the grip, with no likelihood of the flex pulling out of the plug. Note that with this plug, the flex is placed in the cord grip *before* wiring up.

(v) Replace the fuse, checking that it is of the correct current rating for the appliance and its flex. Arrange the wires to rest in the grooves of the plug body and replace the plug cover.

Where the flex is 2-core as fitted to double-insulated appliances and all insulated portable lamps, connect the brown wire to the 'L' terminal, the blue wire to the 'N' terminal and leave the 'E' terminal empty.

Where the flex is non-sheathed and the colour of the two wires are identical either wire can be connected to either the 'L' or the 'N' terminal. Where non-sheathed flex has one wire ribbed or it has a red trace, connect this wire to the 'L' terminal, the other to the 'N' terminal.

The flexible cord of an existing appliance may have the former code colours: red, black and green respectively. Connect the red wire to the 'L' terminal, the black to the 'N' terminal and the green to the 'E' terminal – or replace the flex!

Wiring a non-fused plug

The configuration of the three pins and therefore of the terminals is the same as in the fused plug. That is, with the cord grip at the bottom and the cover off, the live 'L' pin is at the right, the neutral 'N' pin is at the left and the earth 'E' pin is at the top. The flex connections are identical and the terminals are either post or clamp type, depending on the make. The clamp type of flex grip is used in all makes, though in the 2A plug the flex is clamped between the plug body and the cover.

Moulded-on 13A fused plugs

Electrical appliances are now being sold fitted with moulded-on 13A fused plugs. The plugs cannot be rewired. Access to the plug fuse is from the face of the plug.

Fused connection units

Some small fixed appliances – such as extractor fans, cooker hoods, under-cupboard kitchen lighting and waste disposal units – have a flex, but this is connected permanently to a *fused connection unit* rather than to a plug and socket. See page 122 for details of the different types and how they are wired up.

3: DIY housewiring

Housewiring comprises that part of the home installation where the current-carrying cables are fixed permanently in position and which is described as fixed wiring.

The cables are run into ceiling roses, other light fittings, socket outlets, switches, fused connection units and other wiring accessories. Most of the wiring accessories are mounted on boxes which are either moulded plastic or metal. The boxes are of different shapes and sizes to fit the appropriate accessories. In all instances the cable, usually PVC-sheathed, passes into the mounting box and the individual conductors (wires) are connected to the accessory terminals. A few inches of sheathing are removed from the cable to expose the wires and it is a requirement of the regulations that the unsheathed portion of every cable is contained in an enclosure of non-combustible material. With most accessories the enclosure comprises the accessory and its mounting box. Some wiring accessories, which include modern ceiling roses, junction boxes and consumer units, are totally enclosed or have an integral backplate. These accessories do not require boxes and are fixed direct to walls, ceilings or other parts of the house structure.

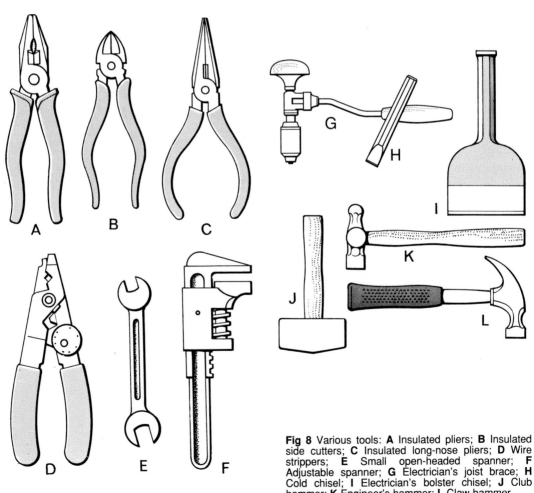

Fig 8 Various tools: A Insulated pliers; B Insulated side cutters; C Insulated long-nose pliers; D Wire strippers; E Small open-headed spanner; F Adjustable spanner; G Electrician's joist brace; H Cold chisel; I Electrician's bolster chisel; J Club hammer; K Engineer's hammer; L Claw hammer.

DIY housewiring

Where boxes are required these can be mounted on the surface or, if metal and of the flush type, sunk into walls. As already stated, much of the work in housewiring is non-electrical and the tools used are those found in most householders' DIY kits but some special tools are recommended and make the work easier (see Figs 8 and 9).

Housewiring cables

Cables used in housewiring today are almost exclusively flat PVC-sheathed (see Fig 10A). This type of cable consists of two, or three, copper conductors enclosed in insulation of different colours, together with an uninsulated copper earth conductor, the whole enclosed in an outer PVC-sheathing. Most of the circuit wiring is carried out using the 2-core plus earth version, one conductor sleeve being red, the other black. The red conductor is used in the live side of a circuit, the black is used in the neutral side; however, sometimes the latter is used in the live side in which case the ends need to be identified by red PVC-sleeving or adhesive tape. The black wire in the live side is usually that from the switch to a light of a lighting circuit and is termed the 'switch return wire' or 'switch-wire'. The 3-core plus earth version of the cable (see Fig 10B) is used in multiple switching circuits including 2-way switching. The core colours are red, yellow and blue with bare earth wire. The colours have no significance in home wiring but are used merely to identify the wires running from the switches, all of which are 'live' wires in spite of the colours.

Wherever bare earth wire is exposed – in a mounting box, for example – it must be covered with green/yellow PVC sleeving.

Single-core cables

There are two types of single-core cables used in house wiring: (i) insulated non-sheathed; and (ii) insulated and sheathed.

The non-sheathed cable (see Fig 10C) is

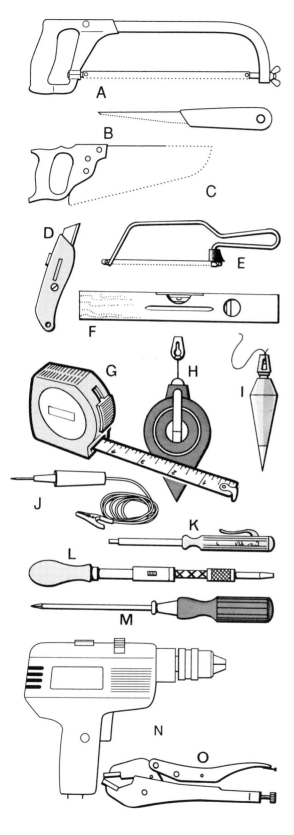

Fig 9 Various tools: **A** Adjustable hacksaw; **B** Padsaw; **C** Floorboard saw; **D** Trimming knife; **E** Junior hacksaw; **F** Spirit level; **G** Steel tape; **H** Chalk line reel; **I** Plumb bob and line; **J** Continuity tester; **K** Neon screwdriver tester; **L** Ratchet screwdriver; **M** Cross-head screwdriver; **N** Electric drill; **O** 'Mole' wrench

30 DIY housewiring

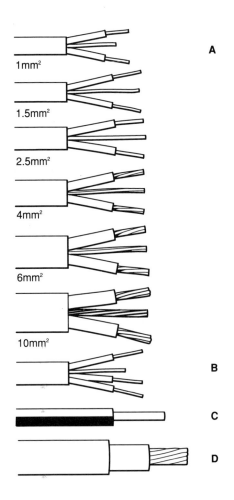

Fig 10 Cables used in house wiring: **A** 2-core and earth PVC-sheathed; **B** 3-core and earth PVC-sheathed; **C** Single-core insulated; **D** Single-core insulated and sheathed.

used as an earth continuity conductor and as an earth bonding lead and has green/yellow PVC insulation. The cable does not normally have to be enclosed in conduit or trunking as it does not carry electric current except when there is a leakage to earth (see Earthing and bonding, page 185).

In older wiring installations, unsheathed cable was used in conduit, but these days any cable used in conduit in the home is normal PVC-sheathed and insulated.

Single-core insulated and sheathed cable (Fig 10D) is mainly used for connecting the consumer unit to the electricity board's meter and from the meter to the service fuse and neutral terminal block.

Cable sizes

The sizes (cross-sectional areas) of the current-carrying copper conductors range from $1.0mm^2$ for a lighting circuit to $10mm^2$ for a cooker circuit and $16mm^2$ (or larger) for the meter leads (see Table 4, page 37).

The conductors of the three smaller-size cables ($1.0mm^2$, $1.5mm^2$ and $2.5mm^2$) are single-strand solid conductors. Those of larger sizes are 7-strand conductors.

Cutting and stripping cable

Cable can be cut using a good pair of side cutters, though the larger sizes may need more than one bite. To strip off the PVC-sheathing, run a sharp trimming knife down the centre of the cable (over the earth wire) and then peel back the sheathing before cutting it off with the side cutters. The insulation on the conductors is then stripped off with wire strippers.

Installing cables

PVC-sheathed cables may be run on the surface of walls and ceilings inside the house, through walls and other structures, in ceiling voids, under floors and in the roof space.

Cables under floors

The first task is to raise the necessary floorboards (see Fig 11). Where a cable is run in a ceiling void beneath floorboards it may rest on the ceiling or other structure. Where it has to cross joists it must be threaded through holes drilled in the joists at least 50mm below the tops of the joists and preferably in the middle. The traditional tool for this is an electrician's brace (see Fig 8) but many electricians these days will use an electric drill or a cordless drill.

Cutting notches in the joists is not good practice, though it may be inevitable close to a wall. Where notches have to be cut, protect the cable from damage (from floorboard nails) by fitting a metal plate over it. Where they are necessary, notches in joists should be positioned under the centre of the corresponding floorboard to reduce the risk of being pierced by fixing nails or screws.

Where laid under a suspended floor the

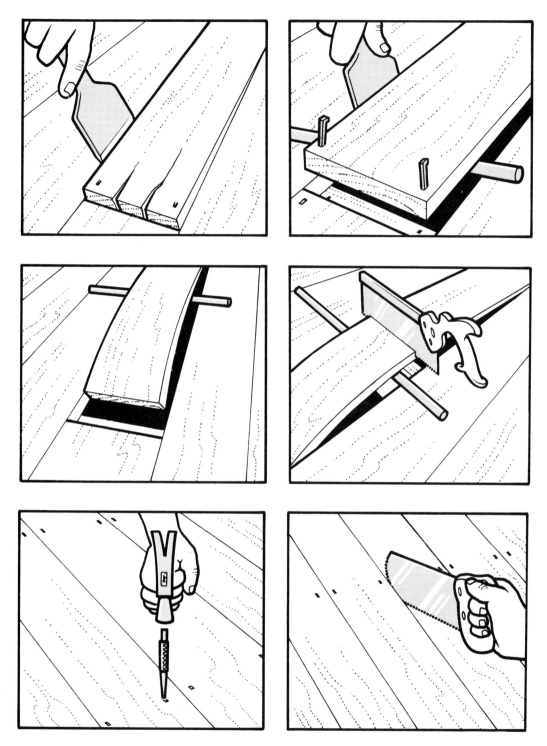

Fig 11 Raising floorboards. *Top left*: Lifting a cut floorboard. *Top right*: Dowel or cold chisel is placed under partly raised board and pushed along as bolster chisel is used to prise the board. *Centre left*: At this stage the board can be 'sprung' by pressure on the end using the dowel as a fulcrum. *Centre right*: An uncut board is sprung and cut at the centre of a joist. *Below left*: Punching nails through a tongued-and-grooved floorboard prior to raising. *Below right*: Cutting the tongue from each side of a tongued-and-grooved floorboard so that it can be raised (a circular saw is a quicker way of doing this).

32 DIY housewiring

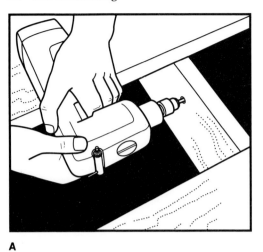

A

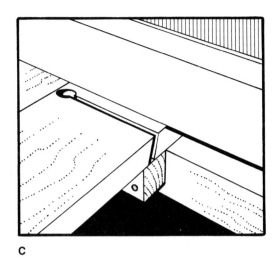

C

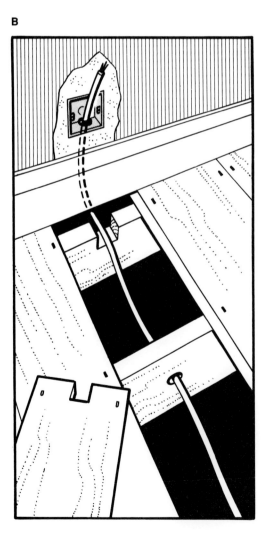

B

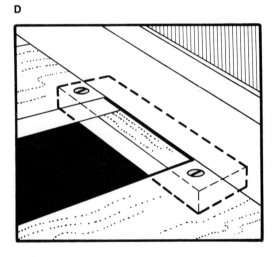

D

Fig 12 Running cables under floorboards: **A** Drilling the joists; **B** The end joist butting the wall is notched, not drilled, and the cable is run up behind the skirting board. The end of the board is notched to clear the cable and a metal plate (not shown) is fitted directly over the position of the notch to prevent damage to the cable; **C** Where a board is cut against a joist under a partition a batten fillet is fixed to the joist to provide fixing for the board, which has been cut at a 45° angle as shown; **D** Where there is no joist under the partition built on the floorboards a batten is fixed to the adjacent boards to support the cut board when relaid.

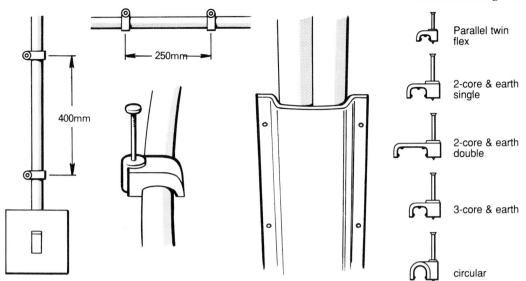

Fig 13 Cable fixings. *Left*: Plastic clips with single-pin fixing with maximum spacings as shown. *Centre*: Steel channelling used to protect cable buried in plaster. *Right*: Various sizes and shapes of cable and flex clip.

cables may rest on the ground or on the building structure without fixings, except to take the weight of the cable where necessary or where the cables are likely to be disturbed or there is access under the floor. In any of those situations the cable is fixed to the joists using cable clips.

Cable clips are sold in different sizes and shapes to suit different types of cable and are available in both single and double versions (for when two cables are run along a surface together). The nails supplied are hardened so that they will go into walls.

Cables run in the roof space

PVC-sheathed cables in the roof space or loft may be run along the joists and rest on the ceiling or other structure. Also they may be run over the tops of joists and fixed to the joists using cable clips, except where the loft is boarded or the cables are likely to be disturbed by persons entering or working in the loft, such as in the vicinity of the cold water storage cistern. In those circumstances the cables are either run through holes drilled in the joists or are enclosed in high impact plastic conduit in those sections where disturbance is likely.

Cable should not be put *under* loft insulation; where lofts have thermal insulation consisting of expanded polystyrene granules as an alternative to fibre blanket or vermiculite, PVC-sheathed cables should be enclosed in conduit where they are likely to be in contact with the insulation as this can attack the PVC, although without significant changes in the insulating properties of the PVC or an increase in the fire risk.

Cables buried in walls

Sheathed cables may be buried in the plaster of a wall or ceiling without the need for additional protection from the risk of mechanical damage, but a simple steel channel (see Fig 13) is a sensible precaution and is not difficult to fit.

To make the chase in the wall, a bolster chisel and club hammer are used, but damage can be kept to a minimum by drilling holes to the correct depth within the marked lines and by scoring along the lines with a sharp trimming knife.

To reduce the risk of damage to cables buried under plaster, the cables should run vertically above and below switches, socket outlets and other accessories since most people will expect cable to be there and will avoid those areas when fixing shelves etc.

Where a horizontal run of cable in the plaster cannot be avoided, the cable should

34 DIY housewiring

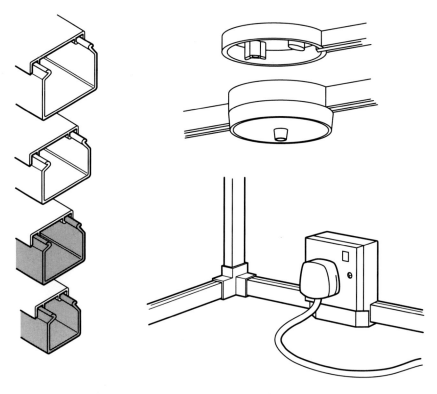

Fig 14 *Left*: Different sizes of white and coloured mini-trunking. *Top*: Ceiling rose adaptor with either one or two lengths of mini-trunking. *Right*: Mini-trunking used to provide power to a socket outlet

be located within 150mm of the ceiling or within a 150mm to 300mm band above floor level, where it is unlikely that wall fixings will be required.

Where socket outlets or switches are to be put in hollow stud partition walls (a timber framework covered in plasterboard or lath-and-plaster), the cable can be run *behind* the plasterboard. You will have to make cut-outs in the plasterboard where the cable meets horizontal 'noggins' (see page 86 for details); you can get special mounting boxes for use in plasterboard.

Cables along walls

Cables fixed to the surfaces of walls (and ceilings) are secured by plastic cable clips having a single steel fixing pin or nail. For vertical cable runs the spacings of the clips should not exceed 400mm; for horizontal runs, 250mm.

Instead of fixing cables to the surface of walls and ceilings they can be run in mini-trunking (see Fig 14) which is neater and eliminates cable clips. Mini-trunking is made from white, high-impact-resistant PVC and has a clip-on cover. It may be fixed with screws or, where surfaces are suitable, with a contact adhesive. Some types are self-adhesive.

There are two ways to use mini-trunking. The first, and least satisfactory, is simply to run the trunking up to normal surface-mounting boxes so that it covers up the cable which passes through one of the knock-outs in the box.

The second, preferred, method is to use the special mounting boxes available (mainly from specialist suppliers, such as electrical wholesalers) where the mini-trunking passes into the box, ensuring there are no gaps. You can get square and rectangular boxes for socket outlets, switches and fused connection units and round boxes for ceiling-mounted switches, ceiling roses and batten holders.

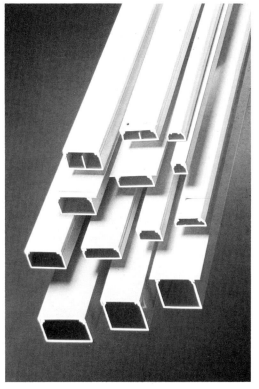

Photo 9 Mini-trunking. *Above left*: Various sizes and shapes of mini-trunking. *Above right*: Mini-trunking used to provide power to surface-mounted socket outlets on a tiled wall in a kitchen. *Below*: Mini-trunking in a garage or workshop protects surface-mounted cable from damage (*Ega*).

Fig 15 *Top*: Cornice trunking can link up with mini-trunking to switches and ceiling roses. *Middle*: Skirting trunking (can also be installed at dado height). *Bottom*: Skirting cover allows cables (or pipes) to be run behind.

Photo 10a A single and a 2-gang switch wired from mini-trunking (*MK*).

The boxes normally need an adaptor to connect them to the mini-trunking.

Mini-trunking is available in various sizes for different sizes (and numbers) of cable and there is a wide range of accessories for joining lengths together, for taking the mini-trunking round corners (or creating a T-junction) and for joining to accessory boxes.

As well as the normal square or rectangular trunking, you can get larger profile *skirting* trunking, which replaces the existing skirting board or can be run horizontally at dado height or vertically next to a door architrave and *cornice* trunking which can be run along the corner between wall and ceiling.

An alternative to skirting trunking is a shaped, veneered-timber, skirting cover which fits over the existing skirting board allowing a space for running cables (or, if you want, central heating or water pipes).

Table 4: **Fixed wiring cables for circuits**

Circuit fuse	Cable size (mm^2)	Circuit (Amps)
Lighting	1 or 1.5	5/6
Immersion heater or other 15A circuits	2.5	15/16
Storage heaters and 20A radial circuits	2.5	20
Ring circuits and spurs	2.5	30/32
30A radial circuits, cookers up to 12kW and showers up to 7.2kW	4 [1]	30/32
Cookers over 12kW and showers up to 10.8kW	6 [2]	40/45
Meter tails	16	Service fuse (electricity company)

[1] only if protected by cartridge fuse or MCB; for rewirable fuse, use 6mm^2 cable
[2] only if protected by cartridge fuse or MCB; for rewirable fuse, use 10mm^2 cable

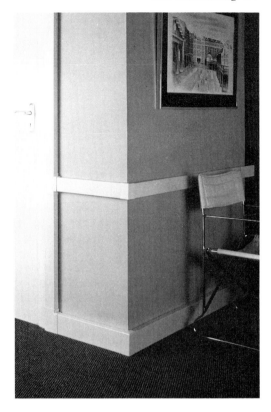

Photo 10b Skirting, architrave and dado trunking systems (*Marshall-Tufflex*).

Photo 10c Cornice trunking used with mini-trunking (*Marshall-Tufflex*).

4: Light fittings

A light fitting is a means of connecting electric lamps to the fixed circuit wiring. The commonest are ceiling-mounted pendant fittings fitted with flexible cord which is connected to the lampholder. Other light fittings are close-mounted ceiling fittings, wall lights, spotlights, track lights, recessed lights and fluorescent lights. Light fittings in general require little if any attention over a long period of years except the periodic replacement of electric light bulbs (lamps) and, less frequently, fluorescent tubes. Over a period of time, lampholders deteriorate and need replacing and the flexible cord (flex) may also need replacing. If you have old-fashioned ceiling roses, you will want to change these for more modern ones.

The different components of a pendant light (ceiling rose, flex and lampholder) are available separately, but many manufacturers these days provide pre-wired pendant sets with a length of heat resisting flex already in place, so that all that needs to be done is to connect it to the fixed wiring. The Crabtree 'Safety Pendant Lampholder' (and 'Batten Lampholder') have a special design which means the pins are not live until a lamp is inserted, which makes changing light bulbs a much safer operation.

Replacing flexible cords

Plain pendants

The plain pendant consists of a lampholder suspended from a ceiling rose by flexible cord. The ceiling rose is basically a junction box for connecting a flexible cord to the fixed circuit wiring. It has a number of terminals or banks of terminals for the various circuit wires and for the flexible cord.

The modern loop-in ceiling rose has three banks of terminals arranged in line, plus an earth terminal. One outer bank consists of three terminals. These are for the neutral feed wires and for the blue core of the flexible cord. The other outer bank consists of two terminals; these are for the switch return wire in the circuit and for the brown core of the flex. The centre bank consists of three terminals, these all being for the live feed wires which are looped in (joined) at this terminal bank. No flex wires are connected to the live terminal bank. The earth terminal of the ceiling rose is used to connect the earth continuity conductors of the circuit cables and to connect the earth core of 3-core flex where this is needed to earth a metal lampholder. Some older lighting circuits may not have an earth conductor – so the earthing terminal in the ceiling rose is not used. But if metal light fittings or lampholders are used, or metal mounting boxes for light switches, the circuit should be earthed with an independent earth conductor from the consumer unit to the ceiling rose. If your lighting wiring is not earthed, it would be worthwhile rewiring using 2-core and earth cable so that it is.

Connecting flex to loop-in ceiling rose

Remove the cover of the ceiling rose and thread in a length of $0.75mm^2$ heat resisting flexible cord. Strip off about 75mm of sheathing from the end of the flex. Strip off about 10mm of insulation from the end of each of the two wires. Connect the brown wire to the outer terminal of the 2-terminal bank. Connect the blue wire to the outer terminal of the neutral terminal bank. Most ceiling roses have post-type terminals. With these, the bared end of the flex wire is doubled back to provide a more effective grip in the terminal. Some older ceiling roses have clamp-type terminals for the flex wires. With these, the bare end of each flex wire is placed clockwise under the clamp screw washer (see Fig 16).

Having connected the flex wires to the terminals, hook each wire over its flex retainer and screw on the ceiling rose cover. Tug the flex to check that the sheathing terminates within the rose. If an unsheathed portion of the flex projects below the cord

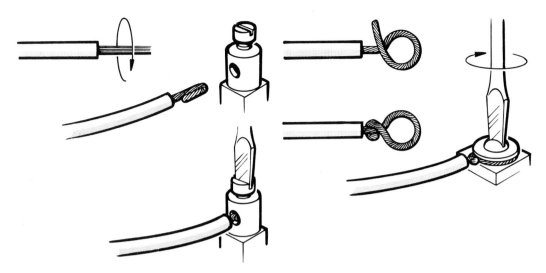

Fig 16 Connecting flex wires. *Left*: To post-type terminals. *Right*: To clamp-type terminals.

outlet of the cover too much sheathing has been stripped off. The remedy is to disconnect the flex, shorten the ends and reconnect the wires to their respective terminals.

Old-type ceiling rose

The old-type ceiling rose has either two or three separate terminal plates which are not situated in-line. Each terminal plate has a terminal – usually a post-type – for the circuit wires and a terminal – usually, but not always, of the clamp type – for the flex wires (see Fig 16).

If you encounter this type of ceiling rose, replace it with a modern one.

Wiring a lampholder

A lampholder is a simple accessory which holds an electric light bulb (or special lamp) and provides electrical contacts to connect the lamp to the circuit. There are two main types of lampholder – the plastic pendant/batten holder type and the screwed metal type, used mainly in table and standard lamps.

Pendant lampholders

Lampholders fitted to plain pendants are made from moulded plastic, including heat

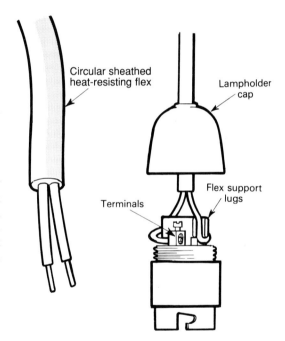

Fig 17 Connecting 2-core heat-resistant flexible cord to a plastic pendant lampholder.

resisting types and types with longer 'skirts' for use in bathrooms. All are designed for use with PVC-sheathed flex.

To connect a lampholder to its flex, use the following method. Cut the flex to the required length. Strip off about 50mm of outer sheathing. Strip off about 9mm of insulation from each of the two wires (brown and blue). Unscrew the lampholder cap and

Light fittings

Photo 11 Plastic lampholders (*Ashley & Rock*).

thread in the flex. Double back the bare end of the brown wire and connect it to one of the post terminals. Do likewise with the blue wire and connect it to the other terminal. Hook the wires over the anchorage slots. Screw on the lampholder cap and check that the unsheathed section of the flex does not protrude from the cap. If the wires are visible they must be disconnected from the lampholder, shortened and reconnected.

Metal lampholders

Metal lampholders are no longer used on pendant lights, but they are found commonly on table lamps and standard lamps where they usually have a screwed thread which goes into the lamp fitting.

Wiring a metal lampholder is very similar to wiring a plastic pendant lampholder, except that the lampholder must be earthed – either by attaching an earth wire to the light fitting itself or by attaching it to the earth terminal on the lampholder. Three-core flex must be used.

Other lampholders

Lampholders fitted to some wall lights and to close-mounted ceiling fittings, have a screwed (threaded) outlet instead of a cord grip and are fixed rigidly to the light fitting. Metal lampholders are usually fitted to metal light fittings. The fitting itself is connected to earth and thus earths the lampholders. Screwed lampholders are also made in moulded plastic versions and are fitted to non-metallic table lamps, floor standard lamps and all-insulated light fittings which are wired with 2-core flex and do not require earthing.

When rewiring a light fitting or portable lamp incorporating a metal lampholder, either earth the lamp or fitting or replace the

Light fittings

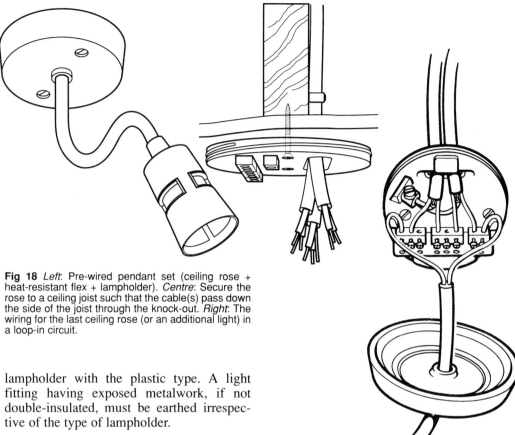

Fig 18 *Left*: Pre-wired pendant set (ceiling rose + heat-resistant flex + lampholder). *Centre*: Secure the rose to a ceiling joist such that the cable(s) pass down the side of the joist through the knock-out. *Right*: The wiring for the last ceiling rose (or an additional light) in a loop-in circuit.

lampholder with the plastic type. A light fitting having exposed metalwork, if not double-insulated, must be earthed irrespective of the type of lampholder.

Replacing a ceiling rose
Whether you are replacing an old-fashioned type of ceiling rose with two or three terminals or a more modern type of rose, the first thing to do, after turning off the power to the lighting circuit (and checking that the circuit is actually dead by trying the wall switch), is to remove the cover of the existing rose and count the number of cables in the rose.

If there is just one, the light is wired on the junction box system; if there are two or three, loop-in wiring has been used and if two it is either an additional light added to the circuit or is the last light on the circuit. Before removing any of the wires, make a note of which wire is connected to which terminal, in particular noting the neutral (connected with the blue flex wire), the switch return (connected with the brown flex wire) and, where present, the live (no connections with the flex). If the switch return wire is black, put a piece of red insulating tape on it. After labelling the cables, disconnect the wires from the terminals and remove the rose.

Wiring the new ceiling rose

The new ceiling rose will be the modern loop-in type having an integral backplate.

It is essential that the new rose has a secure fixing to the ceiling. Ideally, this will be to a joist and the best position for a ceiling rose is overlapping the edge of a joist so that the securing screws go into the joist and the cable comes down the side of the joist to pass through the knock-out in the backplate. If this is not possible, secure a piece of wood between the joists – do not rely on fixings into plasterboard or lath-and-plaster.

Unscrew the cover of the new ceiling rose. Knock out a section of plastic in the base of the rose to take the cables. Thread in the cables and connect the wires to the respective terminals (see Fig 26).

Before connecting the wires it may be necessary to shorten individual wires to reduce any unnecessary slack. Also ensure that the ends of the outer sheathing terminate within the ceiling rose.

42 Light fittings

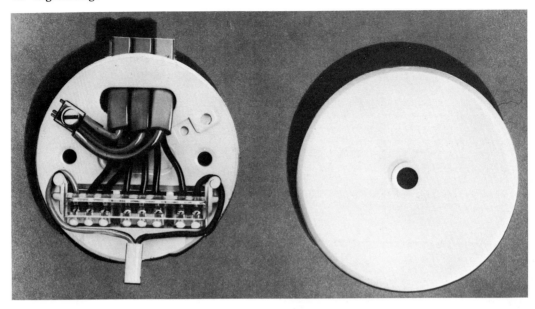

Photo 12 Loop-in ceiling rose with wires connected (*Ashley & Rock*).

Photo 13 As an unsheathed portion of flex protrudes out of ceiling rose, the wires must be shortened and reconnected (*Ashley & Rock*).

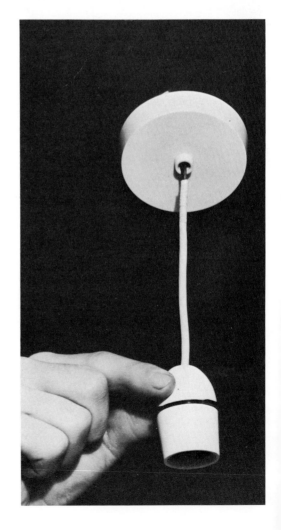

Having trimmed the wires as necessary, connect the single switch return wire (a red wire or a black wire wrapped in red PVC adhesive tape) to the inner terminal of the two-terminal bank. Connect the black neutral wire to a terminal in the outer terminal bank. Where there are two or more neutral wires, separate them and connect them to individual terminals in the bank. Where there are additional live (red) wires, connect these to the central terminal bank.

Fix the ceiling rose, using countersunk screws. Connect the 2-core PVC-sheathed flex to the flex terminals. Connect the moulded plastic lampholder. Re-fit the shade, insert the bulb and turn on the power.

With a pre-wired pendant set, only the fixed cables need to be connected.

Installing a decorative light fitting

When replacing a plain pendant by a decorative type of light fitting the procedure is much the same as for replacing an old ceiling rose. But there is one important difference. A decorative lighting pendant has a ceiling

Light fittings 43

Photo 14 'Klik' ceiling fitting used (*left*) for a plain pendant light and (*right*) for a decorative pendant light fitting (*Ashley & Rock*).

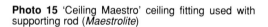

Photo 15 'Ceiling Maestro' ceiling fitting used with supporting rod (*Maestrolite*)

plate which, unlike a conventional ceiling rose, contains no terminals. Instead the flex wires of the fitting terminate in a cable connector and the circuit wires protruding from the ceiling are connected to this. The cable connector is usually a 3-terminal type, but where the light has an earth terminal, the 2-terminal type can be used.

The ceiling plate of some pendant fittings is not enclosed and has no backplate. As the regulations require that the flex, the cable connector and the ends of the circuit wires from which the sheathing is removed must all be contained in an enclosure of incombustible material which, if not the fitting itself, must comprise the ceiling plate and a box or backplate, it is necessary to fix some kind of box or special connector to the ceiling.

There are three choices for doing this: a BESA box, a ceiling pattress and a special ceiling fitting, such as the 'Ceiling Maestro', the 'Klik' fitting or a Luminaire Supporting Coupler (LSC).

Using a BESA box

A BESA box is a metal or plastic box intended for use with conduit and has two screwed lugs tapped for M4 metric screws at 51mm centres which match the holes in the backing

44 Light fittings

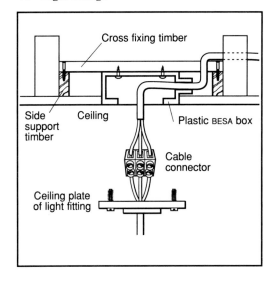
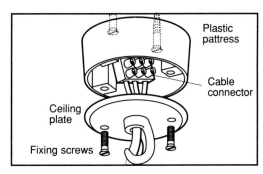
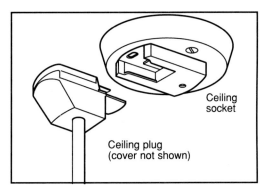

Fig 19 Three ways of supporting a decorative light fitting. *Left*: Using a BESA box mounted flush with the ceiling. *Top right*: Using a ceiling pattress. *Right*: Using a Luminaire Supporting Coupler (LSC).

plate of the light fitting.

Most BESA boxes readily available are made from plastic with a side entry spout. The box must be fitted flush with the ceiling and the easiest way to do this is to cut out the hole for the box in the ceiling with a padsaw, prop a timber offcut against the ceiling from below and then from upstairs with the floorboards raised cut a piece of timber to fit between the joists and mark where the underside of this meets the joists when resting on the top of the box. Fit two short lengths of timber to the sides of the joists and secure the main board on to these with a hole cut in it to pass the cable(s) through.

A single cable can be wired to a three-terminal cable connector as shown in Fig 19. Two or more cables will require a fourth terminal.

Using a ceiling pattress

The round circular pattress designed for use with ceiling-mounted pull-cord switches is ideal for mounting lightweight decorative lights – it, too, has holes 51mm apart. The pattress is fitted to the ceiling in the same way as a ceiling rose and a cable connector is used as for a BESA box.

Using a special ceiling fitting

The advantage of using a special ceiling fitting, such as the 'Klik' fitting, the 'Ceiling Maestro' or the Luminaire Supporting Coupler (LSC) is that these come in two halves, allowing the light fitting to be taken down for cleaning and allowing light fittings to be interchanged. There is usually provision for fitting a screwed brass supporting rod or a hook for a chain.

Although the design of the three fittings is different, the principle is the same: the 'socket' attached to the ceiling takes the fixed wiring and the 'plug' takes the flex attached to the light fitting.

The 'Klik' fitting, made by Ashley & Rock, needs a mounting box on the ceiling (such as a BESA box); the 'Ceiling Maestro' (from Maestrolite) is attached directly to the ceiling and the LSC (made by a number of manufacturers) can be fitted direct or to a box. None of the three systems is interchangeable, but LSCs made by different manufacturers have the same plug/socket configuration. These fittings usually replace ceiling roses, but you can get versions of the 'Klik' and LSC which adapt an existing rose.

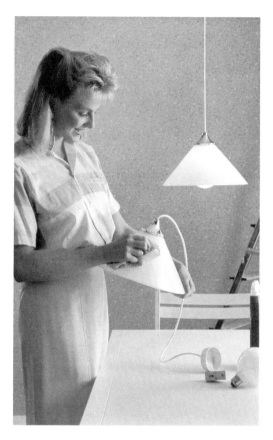

Photo 16 Two of the advantages of using a special ceiling-mounted light connector such as 'Klik'. *Left*: The light can easily be taken down for cleaning. *Right*: Redecorating the ceiling is very much easier (*Ashley & Rock*).

Fixing close-mounted ceiling lights

A close-mounted light fitting is a non-pendant fitting fixed direct to the ceiling. It is made in a wide variety of types but the method of wiring and fixing is much the same for most of them.

The procedure for replacing a plain pendant by a close-mounted ceiling fitting is similar to replacing it with a decorative pendant fitting, the principal difference being that with the close ceiling fitting the circuit cables are usually connected direct to the light fitting.

These light fittings usually have a backplate and need no BESA box sunk into the ceiling. Where the existing ceiling rose has only one cable, this is threaded through the cable entry knockout in the backplate and the wires connected to the fitting terminals. Where there are likely to be excessive temperatures from the lamp in a totally enclosed fitting the wires should be enclosed in heat-resisting sleeves, these usually being supplied with the fitting. If not supplied, such sleeving can be bought from an electrical accessories shop.

Where there is more than one cable at the lighting point with numerous wires including live feed wires it is not possible to run the cables into the fitting. Instead, the cables are drawn back into the ceiling void and connected to a junction box fixed to a piece of timber between the joists. From the junction box a length of heat-resisting cable is run down through the ceiling and into the light fitting where it is connected to the lampholder. The earth wire is connected to an earth terminal in the backplate, or with an all-insulated light fitting the earth wire is terminated in a cable connector. Check the requirements of the chosen fitting.

46 Light fittings

Photo 17 Ceiling fittings. *Left*: Batten lampholder. *Right*: Cord-operated light switch (*Ashley & Rock*).

Fixing and wiring the junction box

Lift a floorboard immediately above the light fitting (unless in an unboarded loft space). Fix a piece of timber about 100mm wide and 25mm thick between the joists at the point where the cables pass through the ceiling to the light. To the board fix a 4-terminal 20 amp junction box (see Fig 29). Pull the circuit cables back into the ceiling void and connect the wires to the terminals of the junction box, taking note of the wires which were connected to the same terminals in the ceiling rose as the flex wires. To the terminals containing these wires connect the two insulated wires of the heat-resisting cable – red to the terminal containing the switch return wire and black to the terminal containing the neutral wire. Connect the earth wire of the cable to the earth terminal.

Replace the cover of the junction box and re-lay the floorboard. Connect the red and black wires of the heat-resisting cable to the lampholder terminals and connect the earth wire to the earth terminal or to a cable connector.

Fitting a batten lampholder

A batten lampholder serves as a complete close-mounted light fitting and can be fixed to the ceiling or to a wall as required. There are are two types to choose from – straight and angled. Both are available with standard or long (HO) skirts (for bathrooms) and come with an integral backing plate.

The procedure for fitting a batten lampholder is much the same as for fitting a ceiling rose: you can get standard mounting blocks for simple (one-cable) wiring and 'in-line' mounting blocks for use with loop-in wiring (up to three cables). Some batten lampholders are pre-wired with heat-resisting tails to connect to the mounting block.

Light fittings

Photo 18 The 'Autoswitch' safety batten lampholder isolates the power once the bulb is removed (*PowerBreaker*).

Replacing a light by a fluorescent fitting

A ceiling-mounted fluorescent fitting can replace any existing light pendant or close-mounted ceiling fitting. No modification to the wiring is necessary, except where the lighting circuit has no earth conductor it is necessary to run one to the fluorescent fitting. This earth conductor consists of 1.5mm^2 single-core green/yellow PVC-insulated cable run from the earth terminal in the consumer unit.

Also, if the present tungsten filament light is controlled by a dimmer switch it is necessary to replace this with a conventional rocker switch.

Where there is already a loop-in ceiling rose for a pendant light, you could leave it in place, mount the fluorescent light alongside it and run a cable up into the roof void and then down again into the light fitting. Alternatively, draw the cables up into the ceiling void and re-connect them in a junction box, running a new cable to the light as described for *Fitting close-mounted ceiling lights* (page 45). With junction box wiring (only one cable), the ceiling rose can simply be removed. For more details on fluorescent lighting, see page 72.

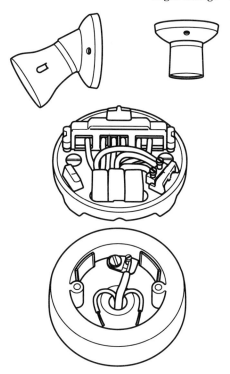

Fig 20 Battenholders and mounting blocks. *Top*: Straight battenholder (*right*); angled battenholder with HO skirt (*left*). *Centre*: 'In-line' mounting block for loop-in wiring. *Bottom*: Standard mounting block.

Installing wall lights

Wall lights can be used to supplement the main lighting in a room or they can be the sole means of lighting. For example, in a large living room a rise-and-fall light can be fixed over the dining table and wall lights fitted in the living area. As a further refinement either or both sets of lighting can be controlled by dimmer switches.

Wiring for wall lights

Wall lights can be connected to an existing lighting circuit, provided the final total number of bulbs of 100W or less does not exceed twelve; alternatively they can be

48 Light fittings

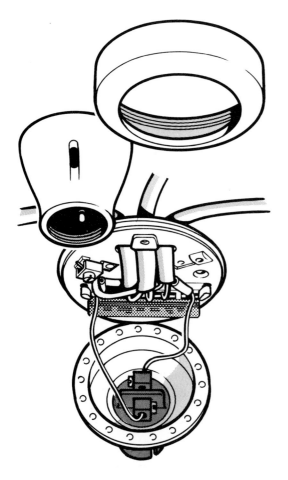

Fig 21 A modern loop-in batten lampholder. The base is identical to that of a ceiling rose of the same make, the wires from the lampholder which can be heat-resistant, are connected to the flex terminals of the base. At the top are shown the lampholder retaining cover and a deep skirt.

supplied from the ring circuit or from an entirely new lighting circuit.

Wall lights can be (a) switched independently of the existing light in the room, (b) switched in conjunction with the existing light, or (c) they can replace the existing ceiling lighting.

Where the wall lights are to be switched independently of the existing room light it is necessary to locate a loop-in ceiling rose or a junction box where the new wiring can be connected to the circuit. If the lighting circuit has been wired on the loop-in system the wall-light wiring can be looped out of the ceiling rose in the room where the wall lights are to be installed (see Figs 22 and 27).

Where the existing circuit is not wired on the loop-in system, but on the junction-box system, the cable feeding the wall-light circuit can be looped out of a junction box (see Fig 31).

Materials required

The materials required for independently switched wall lights are: a length of 1.0mm² 2-core and earth flat PVC-sheathed cable; one 1-way rocker plateswitch; one switch mounting box (flush or surface); a mounting box for each wall light (see page 50); green/yellow PVC sleeving (for covering bare earth wires); one 4-terminal 20A junction box; cable clips; woodscrews and wallplugs.

Planning the wiring

Raise the necessary floorboards, fix the junction box to a piece of timber between two joists (see Fig 29) as central as possible to the wall lights. Drill holes in the appropriate joists to accommodate the cables. Pierce a hole in the ceiling above each wall light position. From the junction box run a length of 2-core and earth PVC-sheathed cable to each wall light, either chased into the wall plaster for flush mounting or run in mini-trunking for surface mounting. Connect the wires in the junction box as shown in Fig 22.

Independently-switched wall lights

Where wall lights are to be switched independently of the existing light in the same room a length of 2-core and earth PVC-sheathed cable is run from the wall light junction box to the loop-in ceiling rose (see Fig 22) or, where the existing ceiling roses are not of the loop-in type, to a junction box (see Fig 31). With this method a separate switch can be installed in any convenient position in the room.

An alternative method is to replace the existing 1-gang switch controlling the existing light with a 2-gang switch. Both the existing light and the wall lights will be switched from the one position but independently as there will be two switches in the 2-gang unit (see Fig 23).

There are two ways you can run the wiring. The simplest is to run a second two-core and earth wire from the junction box to

Light fittings 49

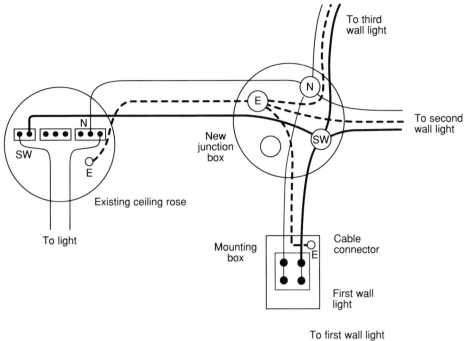

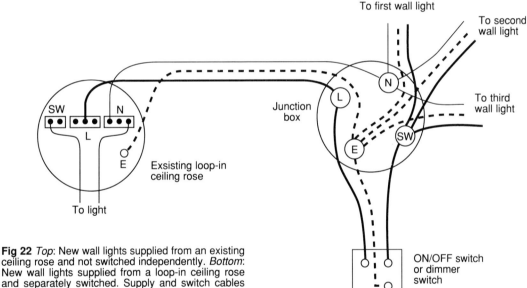

Fig 22 *Top*: New wall lights supplied from an existing ceiling rose and not switched independently. *Bottom*: New wall lights supplied from a loop-in ceiling rose and separately switched. Supply and switch cables not shown.

the existing single-gang switch (neatest if you open up the wall and run it in the same 'chase'). The second is to replace the existing 4-terminal junction box with a 6-terminal junction box and to run 3-core and earth cable to the light switch. In the junction box, the red wire of the 3-core cable is used to take the live supply to the switch and the blue and yellow wires act as separate switch returns for the ceiling light and the wall lights. At the switch, the blue and yellow wires are connected to the L1 (or L2) terminals of the different halves of the switch, the red is connected to the C terminal of one half with a small (red) 'strapping' wire to provide the live supply to the C terminal for the second half.

Wall lights not independently switched

One or more wall lights can be wired directly from the existing ceiling rose, whether you

50 Light fittings

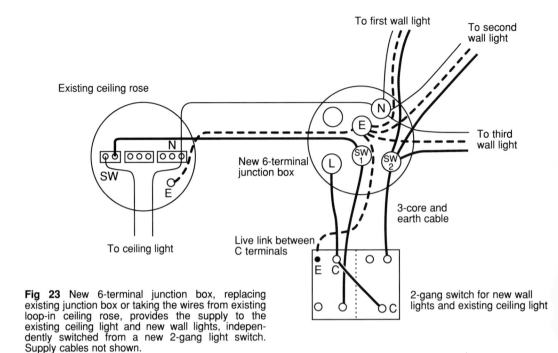

Fig 23 New 6-terminal junction box, replacing existing junction box or taking the wires from existing loop-in ceiling rose, provides the supply to the existing ceiling light and new wall lights, independently switched from a new 2-gang light switch. Supply cables not shown.

have loop-in or junction-box wiring, by connecting the red and black wires of the cable to the terminals containing the flex wires and the earth wire to the earth terminal (see Fig 22).

With this method the wall lights switch on and off with the ceiling light but if you choose wall lights having an integral switch (press button or cord-operated) individual wall lights can be switched on and off as desired.

Replacing the existing ceiling light

Where wall lights are to replace the existing ceiling light and will be the sole fixed lighting in the room, the existing light is removed and the circuit wires are drawn back into the ceiling void. These wires are then connected to a junction box fixed between two joists (see Fig 29). From the junction box a length of 2-core and earth PVC-sheathed cable is run to the wall lights circuit junction box.

Installing wall lights

Wall lights are normally fixed to the wall at a height of approximately 2m above floor level, though the actual height will depend on the height of the ceiling and the type of wall light. Uplighters are usually mounted at a higher level, others lower. It is best to experiment before the lights are wired and fixed.

Most wall lights, including spotlights, have an open baseplate exposing the wires and cable connector. This means that it is necessary to fix a box behind the wall light to contain the unsheathed ends of the circuit wires, the ends of the flex and the cable connector (see Fig 24). The baseplates of many wall lights have two screw-fixing holes 51mm apart and sometimes a cover so that the fixing screws are hidden. One way of mounting these lights is to sink a BESA box into the wall, such that its screw fixing holes (also 51mm apart) match up with those of the light fitting. The BESA box then also provides a space to contain the cable connector.

An alternative – and a method to use where the wall light does not have 51mm screw fixing holes – is to bury a metal architrave switch flush-mounting box in the wall and use this to take the cable connector, fixing the light fitting directly into wallplugs in the wall on either side.

You can also get wall versions of the special ceiling fittings for decorative lights – the 'Klik', 'Ceiling Maestro' and Luminaire Supporting Coupler (LSC) – which may be suitable. The 'Klik' and LSC need mounting

Light fittings 51

Photo 19a Wall-mounted version of Ceiling Maestro with pullcord to operate light (*Maestrolite*).

Photo 19b 'Klik' fitting used to mount wall light (*Ashley & Rock*).

Photo 19c Low-voltage lighting track with integral transformer mounted on a wall (*GE Thorn Lamps*).

52   Light fittings

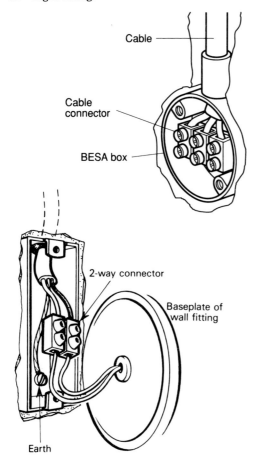

Fig 24 Mounting boxes for wall lights. *Top*: BESA box sunk into wall for lights having fixing screws at 51mm centres. *Bottom*: A flush metal architrave switch box sunk into wall to provide an enclosure for the cable connector. The light fitting itself is secured to wallplugs on either side of the mounting box and an earth wire taken from the mounting box earth terminal to the earth terminal on the light fitting.

boxes sunk into the wall.

Metal wall lights must be earthed, which may mean fitting an earthing terminal.

Wall lights on the ring circuit

Wall lights, as well as other types of lighting, can be supplied from the ring circuit via a fused connection unit (see Fig 72). The fused connection unit is connected to a spur off the ring circuit and is fitted with a 3A fuse. The cable feeding the wall light circuit junction box is connected to the LOAD terminals of the connection unit. The connection unit can be a non-switched version but in many instances a switched version is fitted, the switch being used to turn the lights on and off instead of a conventional switch.

Choosing your lamps

The term electric lamp covers bulbs, strip-lights and fluorescent tubes (see Fig 25).

Most bulbs used in the home are general lighting service (GLS) lamps, which are tungsten-filament bulbs, ranging in sizes from 25W to 150W. The bulbs are made in three principal finishes: clear, pearl and white (opal) though soft opaque colours are also available. Pearl lamps, which have only a slight reduction in light output because the finish is on the inside, are chiefly used in open shades, but where the bulb can be seen a white finish is recommended although its light output is lower. Most GLS lamps have a bayonet cap (BC) fitting.

As well as GLS lamps, you can get a selection of specially shaped lamps – spherical, golf and candle lamps for example – which may have special finishes or a moulded shape. Some of these may have small bayonet cap (SBC) or small Edison screw (SES) fittings.

Many light fittings are designed to take one of a range of internally-silvered reflector lamps, most of which have Edison screw (ES) fittings. The crown-silvered (CS) lamp gives a narrow well-defined beam; the parabolic aluminised reflector (PAR) lamp is used for spotlights and floodlights and is particularly suitable for use in outside light fittings.

Tungsten halogen lamps give a sharp well-defined beam and are available in two versions – single-ended and double-ended (for use in floodlights). They burn very hot and you should never touch the glass.

Fluorescent lamps give five to six times the light output of tungsten lamps of the same wattage and last up to seven times longer. Long tubes are used in fluorescent light fittings; compact fluorescent lamps can replace normal light bulbs and are available in a range of shapes.

The latest type of lamp is low-voltage and runs from a transformer. The lamps themselves are small and unobtrusive and give a clear light (see page 79).

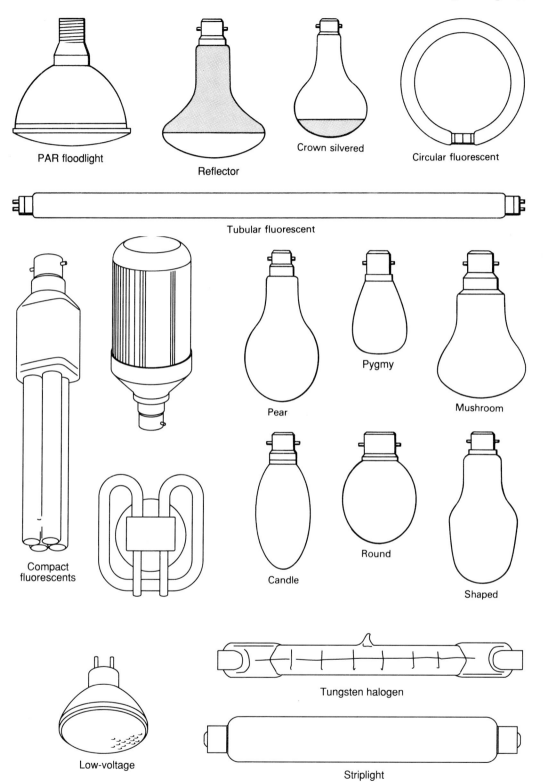

Fig 25 Various types of lamp used in light fittings in and around the home.

5: Lighting circuit projects

A lighting circuit in the home consists of a number of lights and switches in the various rooms and areas. The regulation maximum number of lamps which may be supplied from any one 5A lighting circuit is twelve or the equivalent where some lamps have a wattage above 100W. As explained on page 172, a lighting circuit is wired on either the loop-in ceiling-rose system or on the junction-box system though many circuits are a mixture of both.

The current minimum size of cable for use in lighting circuits is $1.0mm^2$, but many lighting circuits are wired in $1.5mm^2$ cable, which may soon become the minimum size acceptable. The type of cable used is 2-core and earth flat PVC-sheathed cable, though there may still be some older houses without an earth in the lighting circuit or where the lighting is wired in metal ('tin whistle') conduit containing single-core PVC- or rubber-insulated wires.

Adding an extra light and switch

When a new light controlled by its own switch is required, the first job is to locate a source of electricity. This source in a loop-in system will be a ceiling rose. In a junction-box system it will be a junction box located under the floorboards for ground-floor lighting, in the roof space for first-floor lighting or the lighting in a bungalow. Choose the nearest light to the site for the new light for making the connection with the circuit.

To locate a junction box under the floorboards usually means raising one or more boards until one is found. A good place to start looking is vertically above the relevant light switch.

Locating a junction box in the roof space presents no problems. In some installations a central junction box has been used instead of a number of small junction boxes for wiring a lighting circuit on the junction-box system.

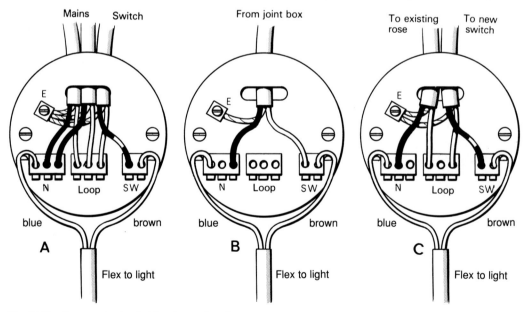

Fig 26 The three methods of wiring the modern loop-in ceiling rose: **A** Three cables with loop-in system; **B** One cable with junction-box system; **C** An additional light added on a loop-in system (the wiring for the last light in the circuit will also look like this). Note that metal light fittings will need an earth wire.

Lighting circuit projects 55

This box will be the source of electricity for the additional light and the new feed cable is run into the box and connected to the mains terminals.

The materials required are: a length of 1.0mm² 2-core and earth flat PVC-sheathed cable; a short length of green/yellow PVC sleeving; a plain pendant consisting of a loop-in ceiling rose and a length of twin PVC circular PVC-sheathed flexible cord; a moulded plastic cord grip bayonet cap lampholder; one 1-way rocker plateswitch; one switch mounting box, which can be either a moulded plastic surface box or a metal flush box.

Instead of a plain pendant the light fitting could be a decorative light fitting, a track light, a fluorescent light or a ceiling light fitting. If the light fitting is anything other than a loop-in fitting, it is better to wire the new light and switch on the junction-box system with only one cable into the lighting point.

Procedure for wiring

Pierce a hole in the ceiling at the point where the new light fitting is to be fixed and pierce another hole in the ceiling immediately above the position of the new light switch. If floorboards are present above the ceiling, raise one above the new lighting point, another above the switch position, and a third above the existing light where the connection for the new light is to be made.

Drill holes in the appropriate joists as necessary to accommodate the new cable running between the existing light and the new light and between the new light and the top of the switch drop where the cable passes through the hole in the ceiling down to the new wall switch.

Take the length of new cable and push the end up through the hole in the ceiling at the new lighting point and with the aid of an assistant draw the cable through to the point above the existing light. Thread the cable through the holes in the joists. At the existing lighting point allow about 300mm for connecting the cable to the ceiling rose below and coil up the end in the void. Return to the site where the new light is to be fixed. Cut the cable leaving about 300mm for connections at the new ceiling rose or batten lampholder. With a ball point pen mark the sheathing MAINS.

Push the end of the remaining cable up through the hole in the ceiling alongside the first cable. Draw this cable up and along the cable route to the hole above the switch position and allow sufficient for running down to the switch and the connections in the switch. Thread this cable through the holes in the joists if necessary and pass the end through the ceiling down to the switch position – either in a wall chase for a flush-mounted switch or mini-trunking for a surface-mounted switch. Make sure that there is sufficient cable for connections within the switch.

At the site of the new ceiling light cut the cable, having allowed sufficient for the connection within the ceiling rose. With a ball point pen mark the sheathing of this cable SWITCH.

Wiring and fixing the ceiling rose

Connect the flexible cord to the ceiling rose and the lampholder to the flex. Knock out the thin section of plastic in the base of the ceiling rose to take the two sheathed cables. Thread in the cables and fix the ceiling rose to the ceiling using two No 8 wood screws. If the rose is not against a joist to which it can be fixed it is necessary first to insert a wood

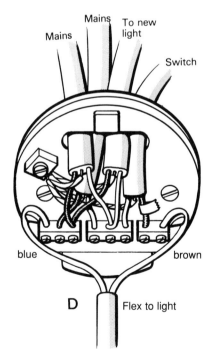

Fig 27 A fourth cable has been connected to this loop-in ceiling rose as a feed cable for an additional light or a number of wall lights.

56 Lighting circuit projects

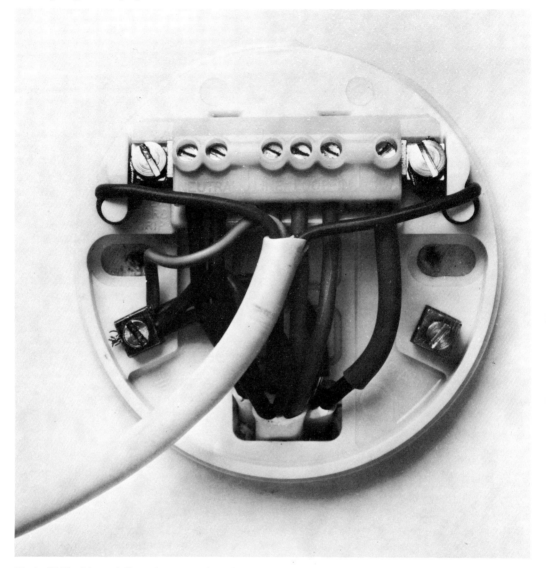

Photo 20 Flexible cord (3-core) connected to a loop-in ceiling rose (*MK*).

batten between the joists.

Strip about 50mm off the end of the cable marked MAINS. Strip about 6mm of insulation from the end of both the red and the black wires. Connect the red wire to one of the terminals in the live central terminal bank (see Fig 26). Connect the black wire to one of the terminals in the neutral bank.

Strip about 50mm off the end of the cable marked SWITCH. Strip about 6mm of insulation from the end of both the red and the black wires. Connect the red wire to one of the remaining terminals in the live central terminal bank. Wrap a piece of red PVC adhesive tape around the end of the black wire and connect the wire to the remaining terminal in the switch return wire bank next to the brown flex wire. Slip green/yellow PVC sleeving over the bare earth wires and connect those wires to the earth terminal of the rose. Screw on the ceiling rose cover.

Wiring and fixing the switch

Lay the cable in mini-trunking and fit a surface-mounted switch box (see Fig 43) or bury it in the plaster and fit a flush-mounted box into the wall (see Fig 44).

Photo 21 Cover of ceiling rose is screwed on to the base which is fixed direct to the ceiling (*Rock/Ashley*).

Connecting new wiring to existing loop-in ceiling rose

Before proceeding turn off the power. Release the loop-in ceiling rose from the ceiling but do not disconnect the wires. Enlarge the cable entry hole in the base of the ceiling rose to accept the new cable. From the floor above push the end of the new cable through the hole in the ceiling alongside the existing cables. Return to the ceiling point and thread the new cable into the ceiling rose. Prepare the end of the cable and connect the red wire to the live terminal bank, the black to the neutral terminal bank and the sleeved earth wire to the earth terminal of the rose (see Fig 27). Refix the ceiling rose and replace its cover.

Junction-box method

Where the new light fitting is to be a special pendant, a close-mounted ceiling fitting or a fluorescent fitting and where there are no facilities for looping in the live feed wires, then the new light with its switch should be wired on the junction-box system. The materials required are the same as for the loop-in system with the addition of a 20A 4-terminal junction box but no ceiling rose as the special light fitting replaces it.

Procedure for wiring

The preliminary work is the same as for the loop-in system, but a junction box is fitted in the ceiling void in a convenient position for wiring and for making the connections within the junction box (see Fig 29). A good place for this is near to the existing ceiling rose or above the proposed new switch or light position.

From the junction box one cable goes to the existing light, another to the new light and the third cable to the wall switch. For the connections at the new light at the junction box see Fig 30. The remaining connections are as described above.

Junction box as electricity source

Where the new light and its switch are supplied from an existing junction box because there is no loop-in ceiling rose, the new wiring is the same as described above but the feed cable is run to an existing junction box (see Fig 31) in the circuit instead of to an existing loop-in ceiling rose.

58 Lighting circuit projects

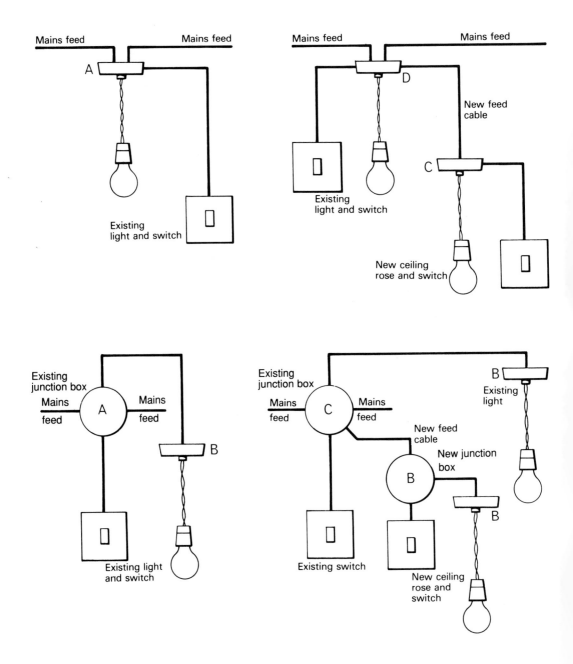

Fig 28 *Above left*: An existing light wired as A in Figure 26. *Above right*: A new light and switch looped out of a loop-in ceiling rose. Connections at the ceiling roses are as Figure 26 and Figure 27. *Below left*: An existing light wired on the junction-box system. The connections at the junction box and ceiling rose are as in Figure 30 and Figure 26. *Below right*: A new light and switch looped out of a junction box. Connections at the junction boxes are as in Figure 30 and Figure 31, and at the ceiling roses as in Figure 26.

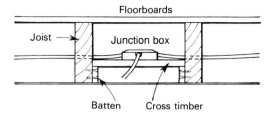

Fig 29 A junction box mounted on a piece of timber fixed between two joists in a ceiling void.

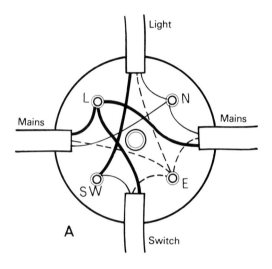

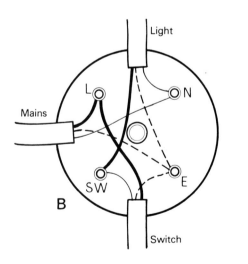

Fig 30 Lighting circuit junction boxes: **A** Connections of wires in a junction box where the lighting circuit is wired on the junction-box system; **B** Connections at a junction box supplying one light and one switch.

Lighting circuit projects 59

Connections at the junction box

Turn off the power and unscrew the cover of the junction box. It will be seen that one of the four terminals contains a number of red insulated wires, another terminal contains a number of black insulated wires and the third terminal contains a red wire and a black wire. Except in older installations, the fourth terminal will contain a number of earth wires.

Strip off about 100mm of outer sheathing from the cable and about 10mm from the ends of both the red and the black wires. Connect the red wire to the terminal containing the red wires; connect the black wire to the terminal containing the black wires. These terminals are the live and the neutral respectively. Slip green/yellow PVC sleeving over the earth wire and connect this wire to the fourth terminal. If the new light fitting is of metal or is a fluorescent fitting it will require earthing, as will metal flush switch mounting boxes. If the junction box has no earth wires connected to it, run a new earth conductor all the way from the main earthing terminal in the main consumer unit.

Wiring a porch light and switch

A porch light can be supplied from the existing lighting circuit (see Fig 32).

The wiring for a porch light is similar to that for an additional light in the house as described above and if there are no loop-in facilities in the conventional or special porch light the new wiring will be on the junction-box system.

The source of electricity for a porch light can be either an existing loop-in ceiling rose or an existing junction box.

First check the hall light. If that has a loop-in ceiling rose the job will be relatively simple. If the hall light does not have a loop-in ceiling rose (or loop-in connections), look at the next nearest light fittings. If none are loop-in lights, it will be necessary to lift a floorboard in the room above or on the landing, or in the roof space if a bungalow, to locate a junction box.

Planning the wiring

The switch for the porch light should be in

Photo 22 A 4-way junction box used on a lighting circuit (*MK*).

Photo 23 PVC twin and earth and PVC-sheathed lighting circuit cables connected to a 4-way junction box (*MK*).

Lighting circuit projects 61

Photo 24 A 6-way junction box used in lighting circuits (*Rock/Ashley*).

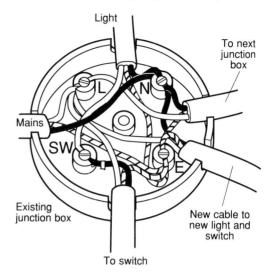

Fig 31 Adding a fifth cable to a lighting circuit junction box where the junction box is used as the supply source for an additional light and switch.

Photo 25a Typical plastic bulkhead light fitting (*Ring*).

Photo 25b Wall-mounted porch light with compact fluorescent lamp (*Coughtrie*).

the hall near the front door. Having decided on its location, pierce a hole in the ceiling immediately above the intended switch position.

The route of the cable into the porch depends on the type of porch light chosen as well as the structure of the building. For a wall-mounted carriage-type lantern or other wall-mounted light the cable from the house should pass through a hole drilled into the wall at the back of the light fitting so that no section of the cable is exposed.

The height of the wall-mounted light will be below ceiling level in the hall and the cable will run down the hall wall from the ceiling void and through the hole in the outside wall. Pierce a hole in the ceiling above the point where the cable will pass through the wall.

Where the porch light is to be a pendant or

62 Lighting circuit projects

Fig 32 Porch and back light extensions. *Top*: A light outside the back door supplied from an existing loop-in ceiling rose and a new junction box from which cables are run to the lights and to an internal (2-way) switch. In addition, 3-core and earth cable is run through the wall to an outside weatherproof 2-way switch. *Bottom*: A wall-mounted porch light wired in the same way.

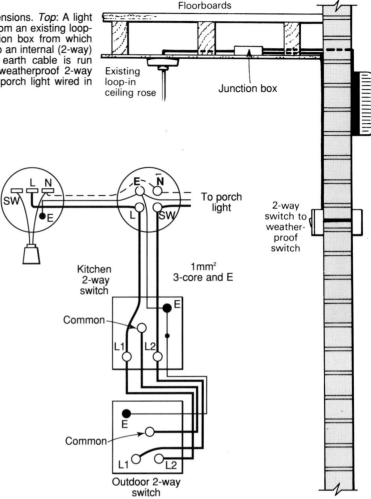

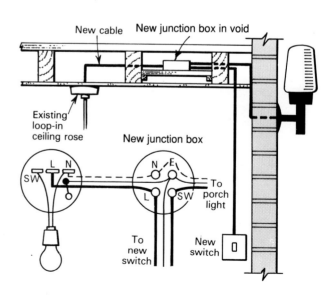

a close-mounted ceiling fitting fixed to the underside of the porch roof, the cable from the house will pass through the outside wall at a higher level than for a wall-mounted light. The point where the cable passes through can be in the hall or in the void above the porch ceiling and therefore under the floorboards in the room above, on the landing, or in the case of a bungalow, in the roof space. First raise a floorboard and check whether it is feasible to make a hole as there may be a joist close up against the wall. On balance it is usually easier to cut the hole below ceiling level in the hall.

Materials required

For either method or type of porch light the basic materials required, in addition to the porch light itself are: a length of 1.0mm^2 2-core and earth flat PVC-sheathed cable; a short length of green/yellow PVC sleeving; one l-way rocker plateswitch and mounting box which can be either a standard switch or an architrave switch and the box either surface or flush fixing; cable clips; wood screws and a 20A 4-terminal plastic junction box.

Procedure for wiring

Drill the hole in the outside wall. Fix the junction box to a piece of wood measuring 100mm by 25mm fixed between two joists in the void above the ceiling near where the cable will drop down the wall to the switch. Take the end of the cable and pass it up through the hole in the ceiling above the switch position. Run the cable to the junction box having first drilled holes in the appropriate joists as necessary to accommodate the cable. Return to the hall and cut the cable, allowing sufficient for the connections at the switch.

Push the end of another length of cable up through the hole in the hall ceiling above the hole in the outside wall. Run this cable to the junction box and with a ball point pen mark it LIGHT.

Return to the hall and measure the length of cable needed to reach the porch light and for the connection. With many fittings, the circuit cables pass through the body of the fitting to the lampholder, so be sure to allow ample cable. Pass the end of this cable through the outside wall into the porch.

From the junction box in the ceiling void run the remainder of the cable to the loop-in rose or existing junction box where the connections to the circuit are to be made. At the junction box mark the end of this cable MAINS. The connections at this junction box can now be made (see Figs 30 and 32).

Wiring and fixing wall switch

The switch (standard width or narrow architrave size) can be mounted either in a surface-mounting box with the cable run down the wall in mini-trunking or in a flush-mounting box with the cable buried in the wall plaster.

Fixing a wall-mounted porch light

Hold the light fitting against the wall in position and mark on the wall the position of the fixings. For other than a circular fitting make sure that the light is straight and upright. Have an assistant to check this, especially for a fitting of the carriage type.

Drill and plug the holes. Examine the end of the fitting. If wires protrude and terminate in a cable connector, this should be a terminal box. If not a box it may be necessary to mount the fitting on to a pattress box. Alternatively it should be possible to remove the wires and run in the wires of the circuit cable having first removed the sheathing for the section within the fitting.

If the fitting has a terminal box, connect the circuit cable to the cable connector. The earth conductor, enclosed in green/yellow sleeving, is connected to the earth terminal of the fitting.

Having wired the fitting (where relevant, in accordance with the makers' instructions) fix it to the wall using brass or other non-rusting wood screws.

Fixing a ceiling-mounted porch light

Wiring and fixing a ceiling-mounted light fitting, whether a pendant or a close-mounted unit, will present no difficulties. If the fitting has an open back it must be mounted on a suitable pattress or a BESA box. With most close-mounted fittings, the sheathed cable passes through a cable entry hole fitted with a grommet and the wires are connected either

Photo 25c Wellglass exterior light, fitted to corner of house (*Ring*).

direct to the lampholder or to a cable connector.

The cable in the porch is fixed to the structure using plastic cable clips.

Connections at the existing point

The final job is to connect the cable to the existing loop-in ceiling rose or junction box. Turn off the electricity at the consumer unit before making the connection.

Installing an outside light

A light installed outside the back door (see Fig 32), or in any outside situation exposed to the rain and weather, must be of a weatherproof type.

There is a wide choice of suitable outdoor lights these days, including traditional and globe lanterns, bulkhead fittings and tungsten halogen floodlights plus a wide range of 'security' lights, fitted with passive infra-red (PIR) sensors which bring the light on when the sensor detects the presence of heat – ie when someone approaches the light. PIR sensors can also be fitted separately to control existing or new lights.

The materials required for wiring an outside light with a switch fixed inside the house – near the back door – are the same as for a porch light, provided the cable from the house passes through a hole in the wall behind the light fitting so that ordinary house wiring cable may be used. Where the cable passes directly into the fitting, it should be sealed with a suitable grommet or bush; where it enters the bottom of the fitting (as with many bulkhead fittings), fit a conduit elbow to the light fitting and direct this into

Lighting circuit projects

the hole in the wall, using an extension length of conduit if necessary. Seal around the conduit with mastic where it enters the wall.

The outside light can be supplied from the lighting circuit with a tapping made at either a loop-in ceiling rose (see Fig 27) or a junction box (see Fig 31) as relevant to the circuit wiring. For the outside light and switch the junction-box system should be used since an outside light fitting has no loop-in facilities.

2-way switching for an outside light

There are many circumstances where it would be especially convenient to be able to switch on the light outside the back door from a switch outside the house. An example is where the garage is at the side of the house and has an access door leading into the area around the back door.

A second switch can be installed almost anywhere, but a sensible and a simple solution is to fix the inside switch near the back door and fix the second switch outside the house immediately behind the inside switch so that the two are back to back (see Fig 32). The outside switch should be weatherproof, but provided it is reasonably protected a splashproof switch such as the MK 'Seal' switch may be used.

The materials required for this job are: one 2-way rocker plateswitch in place of the 1-way switch but using the same mounting box; one 2-way 'Seal' switch; a short length of 1.0mm² 3-core and earth flat PVC-sheathed cable.

Before fixing the box of the switch inside the house, drill a hole through the wall. Thread in the 3-core and earth sheathed cable. If the internal switch is surface-mounted, knock out the thin section of plastic in the base of the mounting box. If it is flush mounted, knock out one of the metal discs from the back of the box and fit a grommet. Thread in the cable and fix the box. Connect the wires to the 2-way switch.

Remove the cover of the 'Seal' switch, fit the cable entry piece in the bottom edge of the casing. Thread in the PVC sheathed cable. Fix the switch unit to the wall and connect the wires to the 2-way switch (see Fig 32).

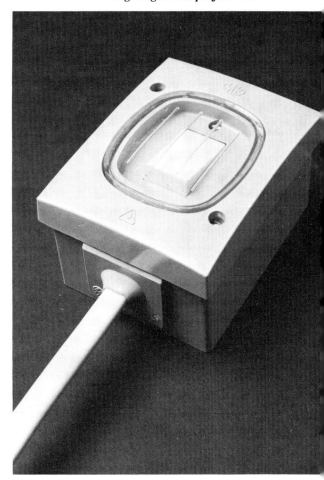

Photo 26 The MK 'Seal' splashproof switch as fixed to the exterior of the house outside the backdoor. The PVC-sheathed cable passes through a special grommet (*MK*).

Outside light from the ring circuit

It is often better to run the cable feeding an outside light and its switch from the ring circuit instead of a lighting circuit.

The cable supplying the light is a spur branching off the ring cable at a convenient position and runs to a switched fused connection unit fixed to the inside surface of the outside wall at the place where the cable passes through the wall to the porch or other outside light. The spur cable (see Fig 72) from the ring circuit to the switched fused connection unit is 2.5mm² 2-core and earth flat PVC-sheathed, and the cable from the switched fused connection unit to the outside light is of the same type but of 1.0mm² size. The fuse in the connection unit should be 3 amp.

Wiring up a security light

A security light can be wired up in exactly the same way as a porch light or a light outside the back door – except that you should take the supply from a ring circuit via a fused connection unit, fitted with a 3A fuse and, preferably, switched so that you can isolate the power to the light.

You have the choice of using a purpose-made security light, with its own built-in passive infra-red (PIR) sensor, or of using a separate sensor which is then wired to one or more new or existing outside lights. The advantage of a purpose-made light is that there is only one connection to be made (unless you use the in-built sensor in the light to power other lights); the advantage of using a separate sensor is that you can position it exactly where you want it – to detect intruders or to welcome you home – which may not necessarily be where you want the light or lights to be.

Security lights and sensors come fitted with a photocell so that the lights are brought on only during the hours of darkness. There is also an adjustment for the length of time the light remains on – typically from a few seconds to several minutes. You can usually set them to be on all the time if you want.

Photo 27a Bulkhead security light with built-in PIR sensor (*Smiths Industries*).

Wiring procedure

To wire a security light you will need 2.5mm² cable for the wiring from the ring circuit to the fused connection unit and 1.0mm² for the wiring from the fused connection unit to the light or the separate sensor.

Where you are running cable along the wall outside, enclose it in rigid PVC conduit to prevent damage and ensure that where cable passes into light fittings or a separate sensor the correct seals are used and the cable forms a drip 'loop' to allow rain to fall off.

Setting the sensor

You can adjust both the sensitivity of the sensor and the zone which it covers. This may take a bit of trial and error to get right – bear in mind that you do not want the sensor to be activated by pets or passing traffic.

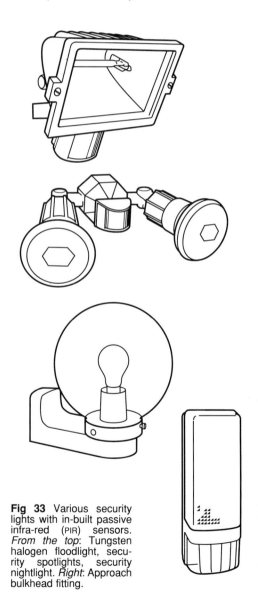

Fig 33 Various security lights with in-built passive infra-red (PIR) sensors. *From the top*: Tungsten halogen floodlight, security spotlights, security nightlight. *Right*: Approach bulkhead fitting.

Lighting circuit projects 67

A   **B**   **C**

Photo 27b Security lighting: **A** Exterior floodlight with 500W tungsten halogen bulb; **B** Separate PIR sensor to bring on exterior light; **C** Exterior tungsten halogen floodlight with built-in PIR sensor (*Mazda*).

Photo 27c Bulkhead security light with built-in PIR sensor. Can switch up to 600W of additional light via slave units (*Home Automation*).

Spotlights

A spotlight produces a beam of light having a shape and intensity determined by the shape of the cowl of the fitting and the type and wattage of the spotlight bulb or lamp (see Fig 34).

There are three basic types of lamp used in spotlight fittings:

(i) An ordinary electric light bulb which emits light through the whole of the bulb surface and provides only general illumination.

(ii) Internally-silvered lamp (ISL), also termed a spot or reflector lamp. This lamp is of special shape and is internally coated with silver or aluminium. The reflector surface gives an accurately controlled beam which is a typical spotlight beam.

(iii) Crown-silvered lamp (CSL) which is a pear-shaped bulb with a silvered crown. It is used in fittings especially designed to reflect back the light. This gives glare-reduced lighting and a tight beam for accentuating

Photo 28 Single spotlight for wall or ceiling mounting (*Philips Lighting*).

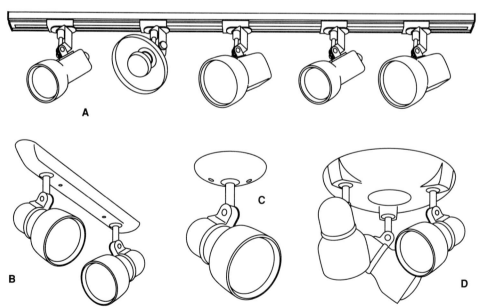

Fig 34 Spotlights used around the home: **A** Track lighting showing different spotlights – *from the left*, medium beam, parabolic reflector with CS bulb, wide beam, medium beam, wide beam; **B** Twin spot for wall or ceiling mounting; **C** Single spot for wall or ceiling mounting; **D** Triple spot cluster for ceiling mounting.

and highlighting features.

These days, an alternative is to use *low-voltage* spotlights – see page 79.

Installing spotlights

A spotlight can be installed as a single light or a twin light, fixed either to the wall or ceiling, a triple spot 'cluster' fixed to the ceiling or two or more spotlights can be fixed to a metal lighting track fixed either to the ceiling or the wall.

Single spotlights

A single spotlight is fixed in a position where the beam of light will illuminate a working surface or highlight a feature in the room. A single fitting, which normally has an open base plate, can be mounted on a round conduit box – a BESA box – which is sunk into the ceiling or the wall depending upon the position of the spotlight. The base plate has two fixing holes spaced at 51mm and is fixed to the box using M4 metric screws in the same way as is a wall light or a pendant light fitting having the same type of base plate. The connections to the circuit wires are made in the BESA box (see Fig 19).

The single light can be switched separately from existing room lighting or it can be switched in conjunction with it. Where three or more spotlights are being installed they should be supplied from a separate circuit.

Twin/triple spotlights

Whilst these may have mounting screw holes the correct spacing for a BESA box, there is usually sufficient room within the fitting to take a cable connector, using screws and wallplugs to mount the fitting.

Track spotlights

A spotlight track is made in 1m lengths. The track is electrified throughout its length, thus enabling spotlights to be positioned anywhere along the track. At the live end of the track is a terminal block to which the circuit cable is connected. Where a track is ceiling-mounted the circuit cable is passed through a hole pierced in the ceiling and into the cable entry at the live end of the track where the wires of the cable are connected to the terminal block (see Fig 35).

Where the spotlight track is to replace an existing ceiling-mounted light (such as a ceiling rose and pendant) the ceiling rose is removed and the live end of the track is positioned in the place of the old light fitting and the circuit cable passed through the cable entry hole. If however there is more than one cable at the existing light (which there will be if the circuit is wired on the loop-in method) it is necessary to terminate the existing cable at a junction box fixed in the ceiling void and run a new cable from the box down into the lighting track. Alternatively, the existing ceiling rose can remain, the track being fixed near the rose and connected to it by a short length of 3-core sheathed flexible cord (see Fig 36).

Wall-mounted spotlight track

For wall mounting, the track may be fixed either vertically or horizontally and fed by a length of 1.0mm^2 2-core and earth flat PVC-sheathed cable running down or up the wall into the live end of the track. As this circuit cable runs into the track itself, its sheathing terminating in the track, no mounting box to contain the wires and cable connector is required as there is with most wall lights and single spotlights. The sheathed cable can either be run in mini-trunking or buried in the plaster. Alternatively, the track can be supplied from a fused plug in a socket outlet using 3-core circular sheathed flex enclosed in mini-trunking.

Track extensions

Where a longer track than the standard 1m is required, extension tracks of 1m in length can be added to the basic starter track. Only the one power source for connection to the basic track is required.

Although a spotlight track can normally be supplied from an existing lighting circuit – especially where the track replaces an existing light – where more than three spotlights are to be fitted a separate circuit is usually required. More than one track will certainly require a new circuit, but first check whether the spotlights will bring the total of lamps on the circuit in excess of twelve of 100W (ie total load 1200W).

70 Lighting circuit projects

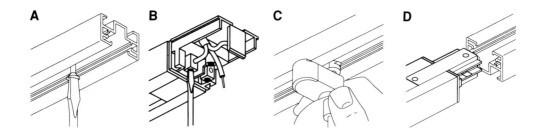

Fig 35 Fitting spotlight track: **A** Screw track to the ceiling; **B** Wire cable into terminal block in 'live end'; **C** Insert adaptor into track and rotate into place; **D** Where extra lengths are required use a joiner piece.

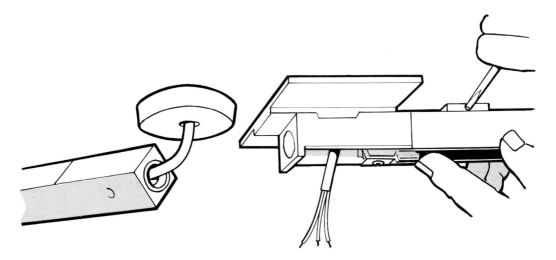

Fig 36 *Left*: Spotlight track connected to an existing ceiling rose with 3-core flexible cord. *Right*: Fixing track to ceiling. The square backplate covers the position of the original ceiling rose.

Fixing the spotlight

A spotlight is inserted into the track at any point and locked in position. Adaptors are used with some spotlight fittings, so when buying spotlights make sure they will fit the track. It is best to buy spotlights of the same make as the track.

Recessed lights

There are two common types of lights used in the home which are recessed or semi-recessed into the ceiling. A downlighter creates a pool of light below it, whilst an eyeball spotlight has a more defined (and adjustable) beam.

Both fit into a hole cut out of the ceiling, with part of the light projecting into the ceiling void.

The best lamp for a recessed light is the internally-silvered (IS) lamp or reflector lamp as it is often termed. The IS lamp reflects the maximum amount of light down on to the surface, which can be the dining table, the kitchen worktop, or any area in a room.

Installing downlighters

When a downlighter is to be recessed or semi-recessed into the ceiling first check that

Lighting circuit projects 71

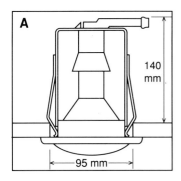 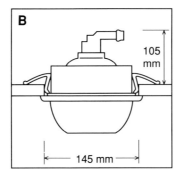 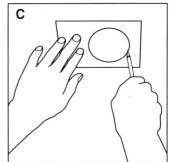

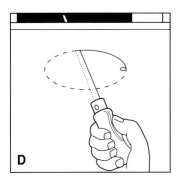 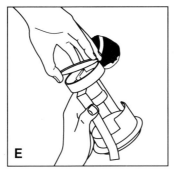 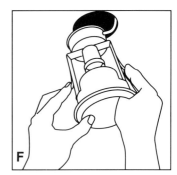

Fig 37 Installing a recessed ceiling light: **A** Typical clearances for a downlighter; **B** Typical clearances for an eyeball spotlight; **C** Use the template provided to mark the hole; **D** Cut through the ceiling with a padsaw; **E** Wire the light following the maker's instructions; **F** Push the light up into the ceiling void and secure.

Photo 29a Track lighting using low-voltage spotlights (*Wickes Building Supplies*).

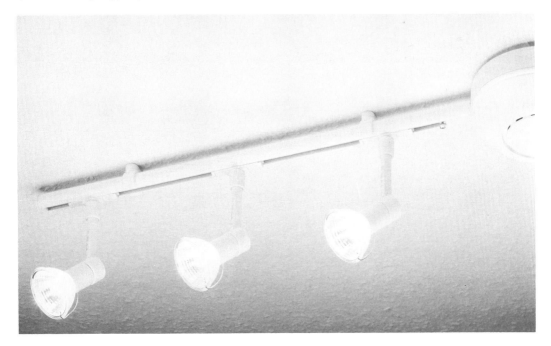

Photo 29b Recessed lights. (*Above*) Swivelling eyeball spotlight; (*left*) Ceiling downlighter (*Philips Lighting*).

there is sufficient height in the ceiling void to accommodate the fitting (see Fig 37). A typical downlighter designed for the home has a depth of only 140mm and it can therefore be accommodated in most ceiling voids, for these usually have a joist depth of 200 or 225mm.

Having decided upon the approximate position for a downlighter, pierce a hole in the ceiling so that it will come in the centre of the downlighter. Raise the floorboard immediately above and check whether the downlighter will clear the joist and that there are no pipes or cables which will obstruct the fitting. If necessary reposition the fitting so that it will clear all possible obstructions. Mark a circle in the ceiling of the required diameter for the downlighter, cut the hole and fix the downlighter in accordance with the maker's instructions. Connect the circuit cable to the downlighter terminals and place in the reflector lamp.

Where a roof space is above the ceiling, such as in a bungalow, there will be no problem in fitting the light but it may be necessary to protect it by fixing timber to enclose the fitting.

Fitting eyeball spotlights

The procedure for fitting eyeball 'spots' is exactly the same as for downlighters, except that you do not have to worry about the clearance above the ceiling. After fitting, swivel the spotlight until it is pointing in the direction you want.

Fluorescent lighting

With their greater efficiency and longer life, fluorescent lights can offer considerable savings compared with the normal tungsten-filament lamps. Most fluorescent lights have a fluorescent tube mounted on or enclosed in a light fitting. The fitting contains starting and current-limiting components which are necessary for the operation of a fluorescent light (see Fig 38). There are three types of fluorescent tube: (i) standard 38mm straight (linear) tubes in sizes from 20W to 125W and in lengths from 600mm to 2400mm, as well as 25mm tubes from 18W to 100W; (ii) circular tubes for use in circular fittings in sizes ranging from 22W to 60W in various diameters; and (iii) miniature tubes in sizes from 4W of 150mm length to 13W of 525mm length. The last are fitted into shaving mirrors, striplights, bulkhead lights and other light fittings requiring a short tube of low wattage

Compact fluorescent lamps (see page 52) are designed to be fitted in place of ordinary tungsten light bulbs. They offer considerable savings in electricity and replacement costs, but there is usually a noticeable delay before they come on.

Installing a fluorescent tube

A fluorescent light can be fitted in the place of any tungsten-filament light using the same circuit wiring. The only likely required modifications are to provide earthing at the fitting and to replace a dimmer switch by a rocker switch, as an ordinary fluorescent fitting cannot be controlled by a dimmer switch.

How a fluorescent light works

A fluorescent tube consists of a glass tube coated on the inside with fluorescent powder and containing argon gas. At each end of the tube is an electrode (or cathode) connected to a bi-pin lamp cap which fits into a bi-pin lampholder.

When the lamp is switched on a stream of electrons flow between the electrodes, bombard the glass tube, excite the fluorescent powder causing it to fluoresce and produce the characteristic light. Initially the flow of electrons has to be started. This is done by pre-heating the electrodes by means of either a switch starter or a transformer.

In the switch-start method (see Fig 38) the circuit contains a starter switch in a small canister (see Fig 39). When the light is switched on, contacts in the switch close and after a short interval during which time the

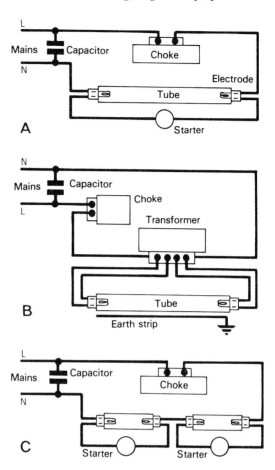

Fig 38 Fluorescent light-fitting circuits. *Top*: Switch-start circuit. *Centre*: Switchless or 'Quickstart' circuit. *Bottom*: Twin-tube circuit with two 20W tubes wired in series using the one choke.

electrodes are heated, the contacts open causing a surge of very high voltage from a choke which is discharged between the electrodes and initiates the electron flow. The choke then acts as a current limiter. Canister switches for switch-start fittings are removable and when one fails it can be readily replaced. Slimmer krypton-filled tubes are recommended for use in switch-start fittings.

In the other method, termed the switchless or 'quickstart' system, a fluorescent tube containing an earthed metal strip is used to assist the starting. The transformer has low-voltage tappings to supply the electrodes in order to heat them, but once the tube is operating the transformer winding serves as the current-limiting choke. It is therefore necessary to provide earthing to the quickstart tube in order to start it.

74 Lighting circuit projects

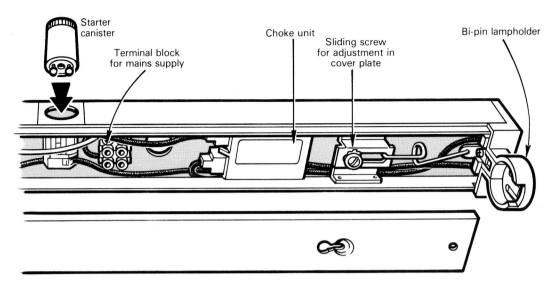

Fig 39 Interior view of a fluorescent light fitting of the switch-start type showing the components and terminal block for the circuit wires.

Photo 30a Straight fluorescent lights (*Linolite*).

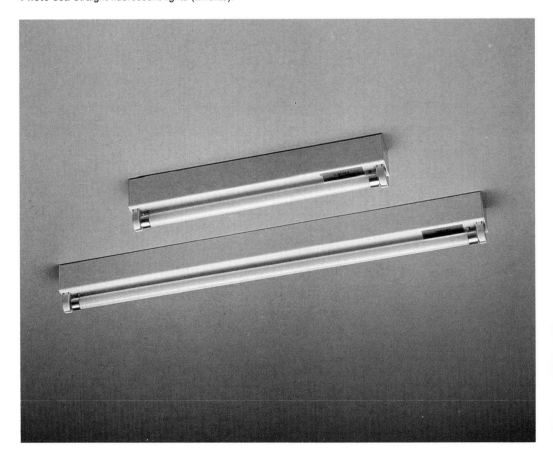

Photo 30b *(above)* Fluorescent lighting in a kitchen (*Linolite*).

Photo 30c *(below)* Undercupboard fluorescent strip-lights provide concealed lighting (*Linolite*).`

Lighting circuit projects

Replacing tubes

A fluorescent tube has an expected average life of 7000 hours as compared with 1000 hours for an ordinary electric light bulb. This is equivalent to many years of life so replacing a tube will be after a long interval. For a quickstart circuit a replacement tube must be of the quickstart type or it will not start. For switch-start circuit either a normal tube or a quickstart tube can be fitted as a replacement as both will operate equally well.

Colours of fluorescent tubes

Of the various tones and 'colours' of fluorescent lighting, de-luxe warmwhite, warmwhite and other warmer colours are most suitable in the home. Some of the warmer tones are given appropriate trade names such as 'Warmtone' and 'Softone'.

Maintenance

When a fluorescent tube reaches the end of its useful life it rarely fails completely as does a tungsten-filament bulb. Neither does a faulty tube fail completely at first. The various faults including flaws on the components cause the light from the tube to behave in a characteristic manner from which the fault can be diagnosed. See Table 5, right.

Illuminated ceilings

An illuminated ceiling consists of aluminium strip section fixed in a grid form to support light-transmitting plastic panels. The panels are made in various colours and patterns to suit individual tastes. Most panels measure 600mm × 600mm but other sizes are available. Suspended ceilings including lights and the wires are available in kit form, which is the best way of buying them.

Lighting for the ceiling

The lighting for an illuminated ceiling is provided by two or more fluorescent tubes (see Fig 41). The tubes can be in batten-type fittings which are complete and contain the necessary starter, choke and other components. Alternatively, plain tubes can be used in conjunction with separate chokes and starters. The tubes are fixed to the ceiling with Terry clips. They are fitted with push-on shrouded lampholders and connected by fly-leads to pre-wired ballast and components housed in a ceiling-mounted casing.

Table 5: **Fluorescent tube failures**

TUBE BEHAVIOUR	POSSIBLE CAUSE
Tube appears dead.	Circuit fuse blown, faulty lampholder, broken tube electrode, break in circuit wiring
Electrodes glow but tube makes no attempt to start.	*Switch-start type*: If glow is white, faulty starter; if red tube is at end of its useful life. *Quick-start type*: Wrong type of tube; ineffective earth connection.
Tube glows at one end when trying unsuccessfully to start.	*Switch-start type*: Lamp holder at dead end of tube short circuited. *Quick-start type*: Broken tube electrode or dead end of tube disconnected.
Tube makes repeated but unsuccessful attempts to start.	Tube very old; faulty starter; low mains voltage.
Tube lights up but at half brightness.	Tube past its useful life.

Ceiling depth

The tubes are fixed to the existing ceiling surface. The translucent panels should be from 150 to 200mm below the tubes for the best effect. Where there is a low existing ceiling, the new ceiling will need to be closer but a depth of 100mm must be allowed between the old and the new ceilings, this distance being sufficient to house the tubes and components.

Quantity of light

The amount of light provided is about 16 watts per square metre. For a ceiling measuring 6ft by 8ft two 40W 1200mm tubes are needed; for an 8ft by 10ft ceiling three 40W 1200mm tubes.

Lighting circuit projects 77

Sections of illuminated ceiling

Where only part of a ceiling is required to be illuminated a simple method is to cut out a section, fix complete fluorescent fittings between the joists and fit a translucent panel in place of the removed section (see Fig 40).

Circuit wiring

Where the power source for the tubes is to be the existing lighting point, the light fitting is removed and in its place is fitted a 4-terminal 20A junction box. From the junction box run a length of 1.5mm^2 2-core and earth flat PVC-sheathed cable to the control box. As the existing switch will be used to control the illuminated ceiling lights no other cable is required.

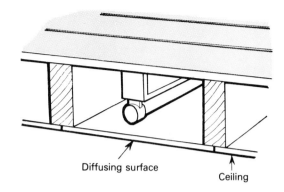

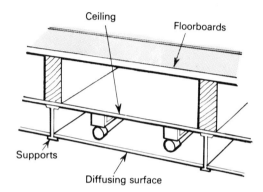

Fig 40 *Top right:* A fluorescent fitting fixed between joists to provide an illuminated section of ceiling, the removed portion of ceiling being replaced by a panel of diffusing plastic. *Right:* Sectional view of an illuminated ceiling fixed to the original ceiling.

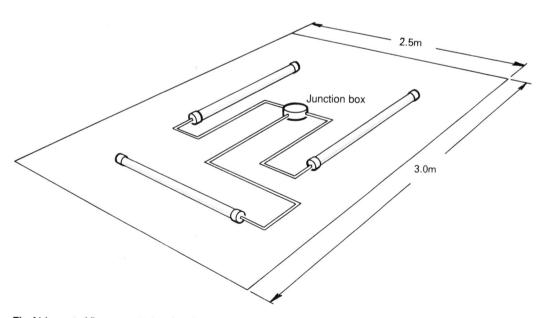

Fig 41 Layout of fluorescent tubes forming an illuminated ceiling and supplied from a junction box fixed in place of the original light, the switch of which is used for controlling the new lighting.

78 Lighting circuit projects

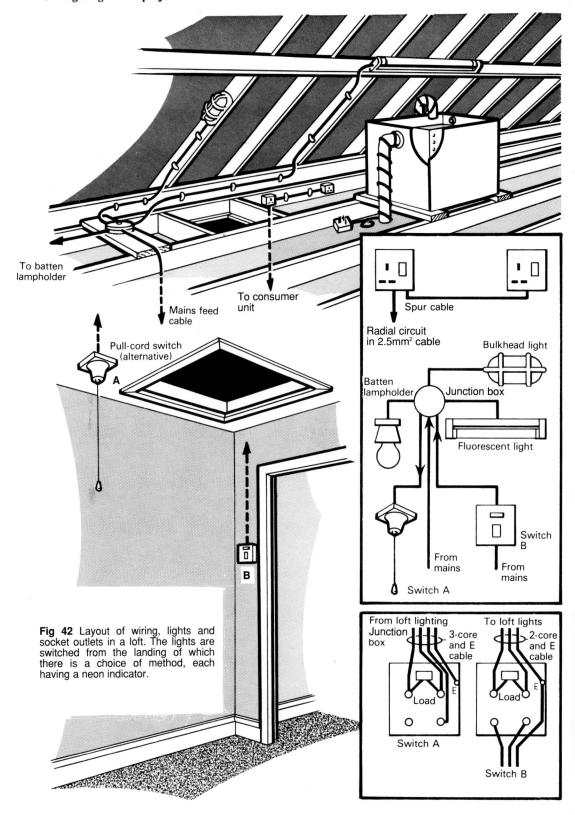

Fig 42 Layout of wiring, lights and socket outlets in a loft. The lights are switched from the landing of which there is a choice of method, each having a neon indicator.

Light and power in the loft

A single light in the loft can conveniently be run off the lighting circuit of the floor below. The connection is made either in a junction box in the roof space or, if the circuit is wired on the loop-in system, from a loop-in ceiling rose, the existing cables to which are in the roof space. The ceiling rose in the room below is released from the ceiling but the wires are not disconnected. Push the feed cable for the loft light through the hole in the ceiling to this rose. Connect the red wire to the centre terminal bank, the black to the neutral terminal bank containing a number of black wires. Slip green/yellow sleeving over the bare earth wire and connect this wire to the earth terminal. Take the cable to a new junction box and from this wire the light and the switch.

Alternatively, and where there are a number of lights, the lighting can be from a separate circuit and the feed cable run down to the consumer unit (see Fig 42). Preferably the loft lighting should be switched from the landing and the switch contain a neon indicator. This can be a 20A double-pole wall switch or a 16A double-pole cord-operated ceiling switch, with neon. Where the lighting is fed from a new circuit the feed cable is 'broken' at the wall switch, but if fed from the upper-floor circuit a 3-core and earth cable is run down to the switch, the third wire being a neutral to energise the neon. Where the switch is inserted into a double-pole feed cable an extra wire is not required. For the power supply 13A switched socket outlets are used. For these a spur cable is run from the ring circuit under the floor in a room below or on the landing using 2.5mm² 2-core and earth PVC-sheathed cable as for other spurs. Remember that there can only be as many spurs on the ring as there are socket outlets and that each spur must feed only one single or one double socket outlet. So if you want to have several sockets in the loft, it would be better to run a separate radial circuit up there in 2.5mm² cable as shown in Fig 42.

Low-voltage lighting

Low-voltage lighting (strictly known as *extra* low-voltage lighting) is familiar enough in shops and offices, but is now available for use in the home.

The lamps run at 12V and low-voltage lighting has the advantages of high efficiency (three times as much light output as normal bulbs), longer bulb life, compact fittings, good colour rendering and a 'cool' beam of light.

The tiny light fittings have tungsten halogen lamps and dichroic reflectors and are generally available as downlighters and spotlights with wide and narrow beam bulbs. A transformer is needed to step down the mains voltage from 240V to 12V. With some light fittings, the transformer is included in the fitting; with others a separate transformer is used.

Fittings with built-in transformers can be used to replace existing fittings (or wired like a new light); those with a separate transformer need much larger sizes of cable between the transformer and the light fittings. The cable needs to be large enough not only to carry the high currents but also to avoid the problems of voltage drop which could affect the efficiency and colour rendering of the lamps. The table shows the maximum lengths of cable which can be used with the common sizes of low-voltage light. The first figure is for a single light wired to its own transformer, the second for lamps wired to transformers designed to supply several lamps.

Table 6: **Lamp cable lengths**

Lamp	Cable size		
	1.0mm²	1.5mm²	2.5mm²
20W	8.5/5.8m	12.8/8.7m	21.1/14.3m
50W	3.5/2.3m	5.1/3.4m	8.4/5.7m
75W	2.2/1.5m	3.4/2.3m	5.6/3.8m

The wiring to a multiple transformer should be on a 5A radial circuit directly from the consumer unit, but a single transformer can be connected into a loop-in ceiling rose or junction box in the lighting circuit in the normal way.

Always keep the low-voltage wiring separate from other mains voltage wiring and ensure there is ventilation space around both light fittings and transformers. Transformers should also be accessible for repair or replacement.

6: Lighting controls

Plateswitches

The modern electric light switch is termed a plateswitch; it consists of a faceplate to which the body of the switch is attached. The faceplate is usually of moulded plastic but is also made in metal of various finishes and even hardwood.

Earlier patterns of plateswitch were dolly switches as were the now obsolete round tumbler switches. All plateswitches now have a rocker action and are termed rocker switches. The rockers are made in various widths and are of piano-key styling.

The plateswitch in common with most wiring accessories has to be mounted on a box. The box may be either of moulded plastic for surface mounting, or a metal box which is sunk into the plaster of the wall, flush with the wall surface.

Most plateswitches installed in the home have square faceplates, with 1, 2, 3 or 4 switches, these being known as 1-gang, 2-gang, 3-gang or 4-gang switches respectively. You can also get rectangular 4-gang and 6-gang switches (which fit into a double mounting box) and a narrow plateswitch – about a third the width of the standard switch. This type is termed an architrave switch as it is designed for fixing into the architraves of doors and in other confined situations as well as in mini-trunking. Architrave switches are made in 1-gang and 2-gang assemblies, the two switches of a 2-gang assembly being vertically in-line on the one faceplate and not side-by-side as are those on the multi-gang square faceplate.

Plateswitches are made in a number of versions for use in the different switching circuits. Those used in the home are 1-way, 2-way and intermediate. 1-way switches are used for switching a light on and off from one position. 2-way switches are used for switching a light on and off from two different positions. Intermediate switches are used where a light is to be switched on and off from three or more positions.

Installing plateswitches

The method of installing a plateswitch depends on the surface of the wall, its construction and whether a surface box or a flush box is to be used.

Height of switch

The normal fixing height for a wall-mounted light switch is around 1.35 to 1.4m from floor level to the centre of the switch, but if preferred a switch may be installed at any other height to suit the householder and the circumstances.

Solid walls

With most solid walls a switch may be either mounted on the surface of the wall using a surface box (see Fig 43) or mounted on a flush box (see Fig 44) sunk into the wall.

Fitting a surface-mounted switch to a solid wall

Fixing a surface box to a solid wall is a simple task. Hold the box against the wall in

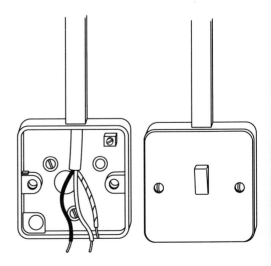

Fig 43 Fixing a surface box to a wall.

Photo 31 (*above*) *Left*: Standard rocker plateswitch. *Right*: Architrave switch (*MK*).

Photo 32 (*below*) *Left*: Surface box for architrave switch. *Right*: 1-gang box for plateswitches (*MK*).

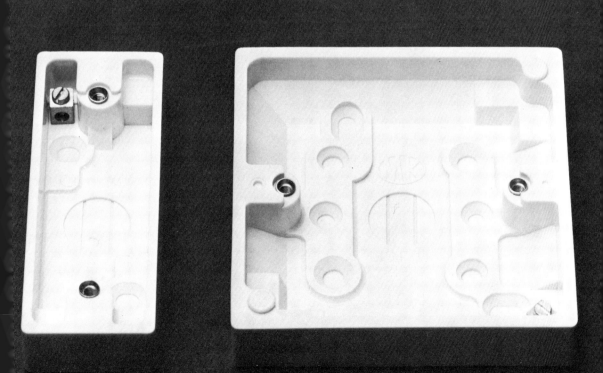

82 Lighting controls

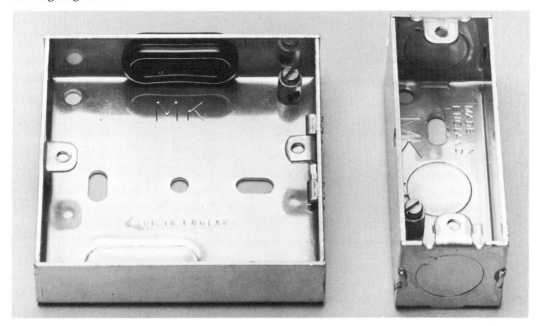

Photo 33 *Left*: 1-gang flush metal box for plateswitches. Box has one adjustable switch fixing lug. *Right*: Flush metal box for architrave switches and also used as backing box for wall lights (*MK*).

Photo 34a Range of plateswitches with hidden fixing screws and screwless terminals (*Crabtree*).

Lighting controls 83

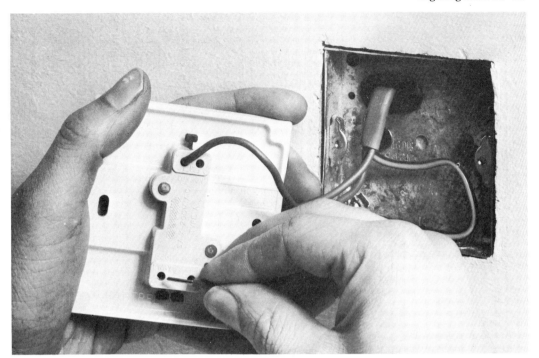

Photo 34b Pushing the circuit wires into the screwless terminals of this pattern of switch (*Crabtree*).

the correct position, using a small spirit level if necessary. Insert the blade of a bradawl through two of the box fixing holes and into the wall. Drill or plug the two holes to a depth of about 38mm to take suitable wallplugs. Knock out a section of the thin plastic from the top edge of the box so that it will accept the sheathed cable running down the wall from the ceiling void. Fix the box to the wall using 30mm No 8 wood screws. Tighten the screws, first ensuring that the box is level. It is much better to use mini-trunking to run cable on the wall surface rather than simply to clip it to the wall. You can get special surface-mounting boxes for use with mini-trunking which have the correct size cut-out in one side.

Fitting a flush box to a solid wall

A plaster-depth box used with a 5A lighting switch has a slotted cable entry in the top edge which is fitted with a PVC grommet to protect the cable sheathing from abrasion, plus knock-outs on one or more of the other sides for use when necessary.

Hold the box in position against the wall using a small spirit level if necessary. Run a pencil around the four edges of the box to mark its position on the wall and draw two parallel lines upwards from the centre of the box for the cable chase. Cut along all the lines with a sharp trimming knife, peel the square and strip of wallpaper off the wall and, with an electrician's bolster (or a sharp cold chisel) and a club hammer, chop out the plaster down to the masonry or brickwork and clean out all dust and debris. Try the box in its hole and, if it projects, chop out some of the masonry until the box sits neatly in the hole firmly against the masonry.

Should the hole be deeper than the depth of the box or should it not have been possible to keep the surface flat, this can be remedied by using wall filler. Mix the required quantity of filler. Push a wet sponge into the hole or use a paintbrush to wet the masonry. Line the bottom of the hole with the filler and push in the switch box. Use the end of the hammer handle to tap the box in until it is flush with the wall along all four sides. Any surplus filler will, if wet enough, ooze out between the box and the wall. Leave the box in position for a few minutes, then remove it leaving the filler to harden.

The box is placed into its hole and the

84 Lighting controls

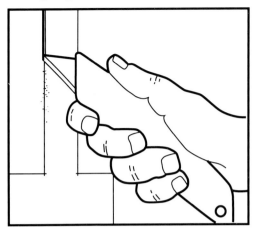

A Draw lines on the wall for the mounting box and the cable 'chase' and cut through the wallpaper with a trimming knife

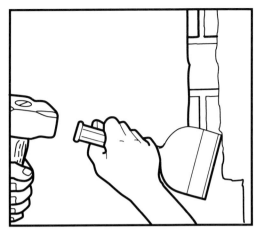

B Chop out the plaster with an electrician's bolster

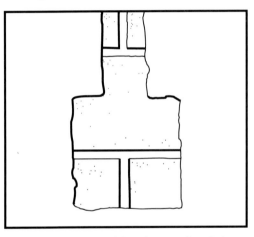

C Clear out dust and debris to leave a clean hole

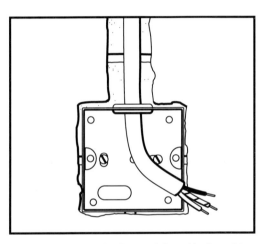

D Secure the mounting box and thread in the cable

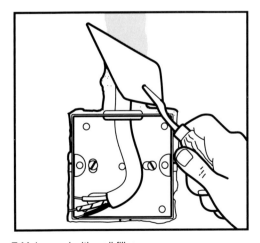

E Make good with wall filler

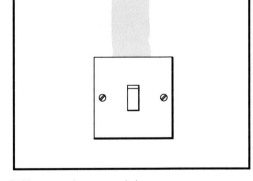

F Wire up and secure switch.

Fig 44 Fitting a flush-mounted light switch

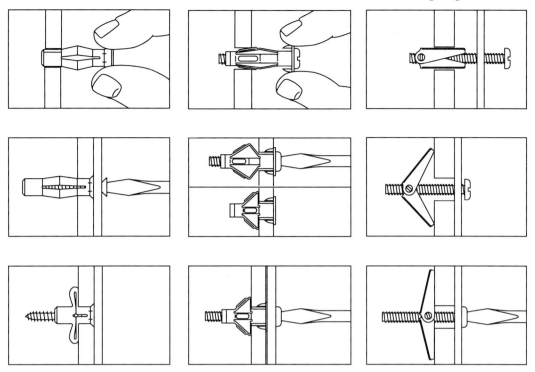

Fig 45 Three types of hollow wall fixing suitable for surface-mounted electrical accessories. *Left*: Hollow wall plug (use like a normal wall plug). *Centre*: 'Interset' fixing ('set' fixing before fixing accessory). *Right*: Spring toggle (fit accessory with fixing – note toggle is lost if fixing removed).

fixing holes marked with a pencil. Drill the holes and fit wallplugs suitable for No 8 screws. Thread the end of the cable through the cable-entry hole and fix the box using No 8 round-head screws. Make good the wall with filler before wiring and fitting the switch.

Dry partition walls

Either a surface box or a flush box can be fixed to a dry partition wall – timber framing covered with plasterboard sheets.

Fixing a surface box to a dry partition wall

A surface box is fixed to a dry partition wall with wood screws, but instead of using normal solid wall plugs, special *hollow* wall fixings are used (see Fig 45).

Hold the mounting box in position against the wall and mark the positions of the fixing holes. Drill holes to accept the fixings (check with the fixing instructions for the size these need to be) and fit the fixings to the walls.

With hollow wall plugs, the fixing is put into the wall first; with 'toggle' fixings, the fixing and mounting box are fitted at the same time – and note that the box cannot be removed without losing the fixing inside the wall.

With cable run on the wall surface in mini-trunking, use the special box available which has a hole in one side to accept the mini-trunking (and the cable inside it); where the cable is run down inside the partition wall and through a hole in the plasterboard, remove one of the knock-outs in the base of the box and pass the cable through this.

Fixing a flush box to a dry partition wall

Where the lighting switch is to be mounted on a flush box sunk into a dry partition wall, there are two choices. The first, where the box coincides with a vertical timber stud (or a horizontal timber 'noggin'), is to cut a square out of the plasterboard over the stud or noggin and fit a plaster-depth box directly to the wood – if the box sticks out from the surface, use a wood chisel to recess the timber. The second, where the box does not

86 Lighting controls

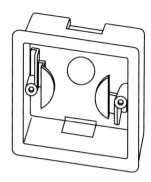

Fig 46a Drylining box for flush mounting electrical accessories in plasterboard.

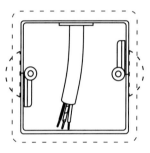

Fig 46b After fitting the box in the wall, the side 'wings' are rotated to hold it in place.

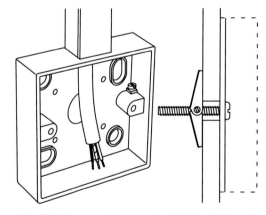

Fig 47 Fixing a surface-mounted light switch. *Left*: The cable is run down the wall in mini-trunking to a special surface-mounting box. *Right*: Toggle fixings used to secure the box to the wall.

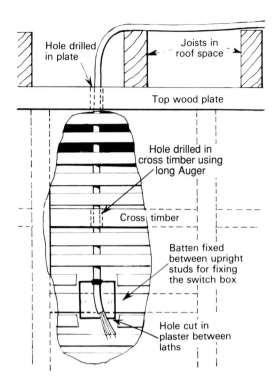

Fig 48 Running a PVC-sheathed cable down inside the cavity of a lath-and-plaster partition.

coincide with a timber member, is to use a special plastic 'dry lining' box, which has 'wings' on the side; both single and double versions are available. A hole is cut out of the plasterboard (using a padsaw) the cable is threaded through the box and the box pushed into place. The 'wings' are then turned into place so that when the switch securing screws are tightened they come up against the back face of the plasterboard to hold the box in place.

Lath and plaster partition

Many old houses have lath and plaster partitions, especially those built before 1914. This type of partition wall consists of timber uprights and cross members covered with laths fixed horizontally, the whole being covered with plaster as were older ceilings before the introduction of plasterboard.

The cable running to a switch fixed to a lath and plaster partition is normally dropped down the cavity of the partition, it usually being necessary to drill two or more timber cross members (see Fig 48).

A switch cannot be fixed to the laths because they split and do not provide firm fixing. It can however be fixed to a timber upright or to a cross member provided either is in a suitable position as described on page 85. You may be lucky with an upright, but cross members are usually positioned too high or too low.

Lighting controls 87

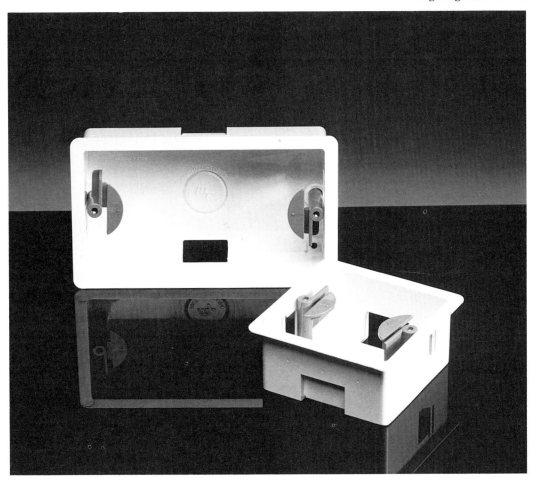

Photo 35 Drylining boxes for mounting flush accessories in plasterboard partition walls (*Marshall Tufflex*).

Where cross members can't be used, it is necessary to provide new fixings in the form of wood battens fixed between two cross members. First chop out the plaster covering two adjacent laths and remove the laths; this will mean cutting them at the uprights. Cut a batten to fit *across* two uprights, and recess the uprights to the depth of the batten so that it fits flush with the front of the uprights. Make good the wall with a combination of plasterboard and wall filler. With a flush-mounted box, check first that the front of it will lie flush or below the finished wall surface, thread the cable into the box (through a hole drilled in the batten if necessary) and fit the box to the batten before making good. With a surface-mounted box, let the filler dry before fitting the box.

Replacing an old-fashioned light switch with a plateswitch

If you move into an old property which has not been rewired for a long time, you may find the old-fashioned, circular, 'tumbler' light switches mounted on small square wooden pattresses.

Replacing these with modern surface-mounted or flush-mounted plateswitches is not difficult, but you should seriously consider rewiring the whole lighting circuit as it is almost certain that this will still be in the original rubber-insulated single-core cable, the rubber of which will have perished, and that it will not be earthed.

The principles of lighting circuits are covered in detail on pages 172 to 175; here

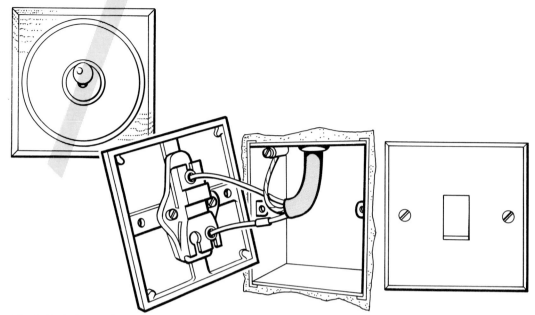

Fig 49 Replacing an old tumbler switch mounted on a pattress with a modern rocker plateswitch mounted on a flush metal box.

we are concerned about the problem of replacing the light switch.

The old circuit is likely to be run in metal conduit buried in the wall plaster and if you are happy to have the new light switch in the same place as the old one, you should be able to use this conduit to draw new PVC-sheathed lighting cable down from the loft space or the ceiling void. For both 1-way and 2-way switching, you will need 2-core and earth cable between the switch and the loop-in ceiling rose or junction box; for 2-way switching you will additionally need 3-core and earth cable between the two 2-way switches.

The presence of the conduit in the wall can affect where you position the new light switch. Surface-mounted switches can go almost exactly where the old switch was positioned (with a hole drilled or pierced in the back of the surface-mounting box to take the cable).

Flush-mounted switches, however, will usually have to be positioned further down the wall, unless you are prepared to cut through the conduit in the wall with a hacksaw blade mounted in a padsaw handle. The reason for this is that the conduit in the wall will finish in the middle of the old switch and the new flush-mounting box will have to be positioned in the wall with its top edge below the end of the conduit unless the box can be placed to one side.

When removing an old type of light switch, you may find that the screws holding the hardwood pattress have rusted in place. If this is the case, split the wooden pattress and remove the screws with a pair of pliers. If there is no protective bush on the end of the conduit, fit one after passing the cable through (fit one at the other end of the conduit as well and both will prevent the cable from chafing on the edge of the conduit).

If you consider the existing wiring is in good enough condition, you can use it to wire up the new plateswitch, but you should use a plastic surface-mounting box if the circuit is not earthed (and not use metal light fittings). A 1-way switch can be replaced with a 1-way plateswitch or a 2-way plateswitch with only two of the three terminals (usually C and L2) being used. With a 2-way switch, note which wires are connected to which terminals of the old switch before connecting them to the new. Old-type 2-way light switches had one terminal on one side of the switch and two on the other. The wire connected to the single terminal goes to the C terminal on the new 2-way plateswitch; the other two wires go to the L1 and L2 terminals.

Lighting controls

Photo 36 A 1-gang and a 2-gang rocker switch mounted on a surface box (*MK*).

Photo 37 Rear views showing terminals of rocker switches. *Top left*: Architrave switch. *Top right*: 1-way switch. *Bottom left*: 2-way switch. *Bottom right*: 2-gang switch, both switches being 2-way as is customary (*MK*).

Replacing a damaged rocker switch

To replace a broken or faulty rocker switch is a simple job. First buy a matching rocker-operated plateswitch which will be either a 1-way or a 2-way version. The same switch can be mounted on the standard 'square' box whether the box is surface or flush fitting. If a surface box it is necessary that the switch matches the box for these have different tones and are of slightly different shapes. If an exactly matching switch for surface mounting cannot be obtained an alternative is to buy a new box with the switch.

1-way switch

Turn off the power. Remove the two fixing screws of the switch and release it from the box. Examine the terminals of the switch. If it is a 1-way type having two terminals, disconnect the two wires but do not disconnect the earth wire connected to the earth terminal in the base of the box, unless it is bare – in which case disconnect it, slip some green/yellow sleeving over the wire and reconnect it to the terminal in the mounting box.

Take the new switch and connect one wire to each terminal. Either wire can go to either terminal if the new switch is a 1-way one (see below for using a 2-way switch for 1-way wiring). Place the wires in the box neatly and fix the switch to the lugs of the mounting box using the original screws.

Do *not* use the new screws which came with the switch as those will have metric threads whereas the existing screws may be 4BA Imperial screws; these are not always interchangeable with the equivalent M3.5 metric screws. An attempt to insert the wrong screw into the brass insert of a plastic box can jam the screw, break the plastic and ruin the box. In a flush-mounting box, trying to fit the original screws can strip the threads on the screw or the box lugs if the screws are the wrong size. Where a new surface-mounting box is being used to match the colour or the shape of the replacement switch, use the screws provided with the switch. To mount the new box, you may be able to use the existing wallplugs in the wall – if not, mark out the fixing holes and drill holes for new wallplugs.

2-way switch

When replacing a 2-way rocker switch the procedure is the same as for a 1-way switch except that the 2-way switch has three terminals and there will be three and sometimes five wires in addition to earth wires.

With the power off, remove the existing switch but before disconnecting any wires note to which terminals they are connected. As the method of wiring a 2-way switching circuit is slightly more complex, it is best to proceed by dealing with one terminal at a time. For example, disconnect the wire from the Common terminal of the existing switch and connect it to the Common terminal of the new 2-way switch. Then transfer the wire or wires from the L1 terminal of the old switch. Finally transfer the wire or wires from the L2 terminal of the old switch to the L2 terminal of the new switch. Place the wires in the box and fix the switch to the box using the existing screws.

Circuits and electrical installation work on 2-way switching is explained in further detail on pages 93 to 96.

2-way switch used as 1-way

A 2-way switch can be used either as a 2-way switch or as a 1-way on/off switch. When, therefore, you remove a broken or faulty rocker switch and find that it has three terminals but the wires are connected to two terminals only, you will know that the switch is 2-way type but has been used as a 1-way. This switch can be replaced by a conventional 1-way plateswitch, the procedure being as already described.

A 2-way rocker plateswitch can replace a 1-way faulty rocker switch where no new 1-way rocker switch is available. The two circuit wires are connected to the Common and the L2 terminals respectively unless stated otherwise on the back of the faceplate. The switch is fixed to the box the correct way up, 'Top' being indicated on the back of the faceplate.

If the second wire were connected to the L1 terminal instead of the L2, the switch would operate as a one-way switch but unless reversed when fixing it to its box it would be 'upside down', the rocker being UP for ON and DOWN for OFF.

Replacing a faulty 2-gang switch

A 2-gang lighting switch consisting of two switches on a 1-gang faceplate usually controls two lights independently. A popular example found in the home is in the hall where one switch controls the hall light and is a 1-way switching circuit (see Fig 53). The other switch is 1 of 2 on a 2-way switching circuit controlling the light on the landing, the second switch being located on the landing.

In this instance although one of the two switches of the 2-gang assembly operates as a 1-way switch and the other operates as a 2-way switch, when buying a 2-gang lighting switch you will find that *both* are in fact 2-way switches.

To replace the faulty switch unit, first turn off the power, remove the two fixing screws and examine the wires in the terminals. In the 1-way switch, there will be a wire in the Common and the L2 terminals, with the L1 terminal blank. All three terminals of the other switch will have wires connected to them. Taking each switch in turn, disconnect the wires and reconnect them to the corresponding terminals of the new switch. When both switches are wired, place the wires in the box and fix the switch assembly to it using the existing screws.

The same procedure is used where both switches are used 2-way or where both are used 1-way. When changing over the various wires double check that the same terminal is used in the new switch as in the old and take care not to separate any jointed wires when taking them out of a terminal. If any jointed wires get separated as you remove them from a terminal make sure they are all properly connected back into the same terminal of the new switch.

Where 2-gang switches are used for 2-way switching of the hall and/or landing lights as described above, it is important to remember that these switches will have live wires inside them unless *both* lighting circuits are turned off at the mains (assuming that the hall and landing lights are on separate circuits).

This is a potential hazard and one which can be avoided if 1-gang 2-way switches are placed one above the other with the top one in each case controlling the landing and the bottom one controlling the hall.

Replacing a wall switch with a ceiling switch

In bathrooms, the light switch should be a cord-operated, ceiling-mounted type rather than a wall-mounted plateswitch. The wiring regulations state that no switch should be within reach of anyone using a bath or shower unless it is a ceiling-mounted type with an insulated cord.

In some older houses, bathrooms may still have wall-mounted plateswitches in the bathroom and a bedroom being converted into a bathroom (perhaps an *en suite* bathroom) will almost certainly have this type. The remedy is to replace the wall-mounted switch by a ceiling-mounted type.

The materials required are: one 6 amp 1-way cord-operated ceiling switch complete with a ceiling mounting block; a short length of $1.0mm^2$ 2-core and earth PVC-sheathed cable; one 4-terminal 20 amp plastic junction box; a few inches of green/yellow PVC sleeving.

Turn off the power and remove the circuit fuse. Remove the wall switch from its box. Disconnect the wires from the switch terminals. Remove the mounting box leaving the cable protruding from the wall. Knock out a little of the plaster around the cable and check that the cable drops down from the ceiling and whether it is enclosed in conduit or whether it is buried direct in the plaster.

Go into the roof space above the bathroom and locate the point where the cable passes through the ceiling and drops down to the switch. If a ground-floor bathroom it will be necessary to lift a floorboard above the bathroom switch.

If the cable dropping down the wall is enclosed in conduit it can be withdrawn and will be long enough to reach the ceiling switch without the need for a junction box and extra cable. Pierce a hole in the ceiling at the point where the ceiling switch is to be fixed. Where possible choose a point next to a joist so the switch can be fixed to the joist. Pass the cable through the hole and return to the bathroom to fit the switch.

If the cable is embedded in the wall and cannot be withdrawn, cut it flush with the ceiling. Fix the junction box to a length of timber fixed between two joists preferably close to the ceiling to protect the junction box. Run the end of the cable into the junc-

92 Lighting controls

Photo 38 *Left*: 1-way 6A ceiling-mounted light switch (also available as 2-way). *Right*: Retractive switch (only operates as long as cord is pulled) with red cord (*Crabtree*).

Photo 39a 16A double-pole ceiling-mounted switch with neon indicator (*Crabtree*).

Photo 39b 40A double-pole ceiling-mounted switch with neon indicator and mechanical indicator (*Crabtree*).

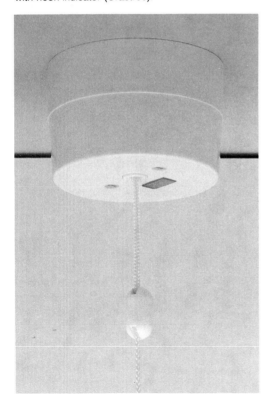

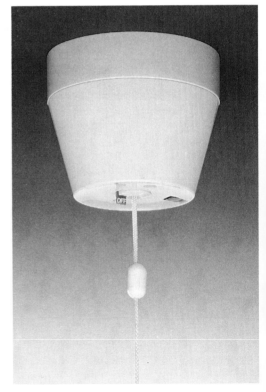

tion box, prepare the end and connect the wires to the junction-box terminals. From the junction box run a length of the new cable down through the hole in the ceiling for connection to the ceiling switch. At the junction-box end connect the wires of the new cable to the same terminals as the old cable. Replace the junction-box cover and return to the bathroom.

Fixing the ceiling switch

Release the ceiling switch from its mounting box by removing the two fixing screws. Knock out a section of the thin plastic covering the cable entry in the base of the box. Thread in the end of the cable protruding from the ceiling. Fix the box to the ceiling joist using No 8 woodscrews. If the cable is not against a joist it will be necessary to fix a wood batten between the joists and drill a hole through it in line with the hole pierced in the ceiling to accommodate the cable.

With the mounting box firmly in position trim the cable to about 200mm in length and strip off about 75mm of the outer sheathing. Strip off about 10mm of insulation from the ends of the red wire and the black wire. Slip green/yellow PVC sleeving over the bare earth wire and connect this wire to the earth terminal in the box. Connect the red wire to one of the two terminals of the switch. Connect the black wire to the other terminal. Place the wires neatly in the box and fix the switch to its box using the two screws supplied. Replace the circuit fuse and turn on the power. Make good the hole left by the old switch – or fit a blanking plate to the old mounting box.

Converting 1-way switching to 2-way

It is often an advantage to be able to turn a light on from two places. This can be done by converting the existing 1-way switching to 2-way. To do this, the 1-way switch is replaced by a 2-way one (unless it is already a 2-way switch), a second 2-way switch is fitted where you want it and the two are linked by 3-core and earth PVC-sheathed cable.

The materials required: two (or one) 6A 2-way rocker plateswitches; two (or one) surface-mounting or plaster depth flush mounting boxes; a length of 1.0mm^2 3-core and earth flat PVC-sheathed cable; a short length of green/yellow PVC sleeving; woodscrews and wallplugs; mini-trunking (for surface-mounted switches).

Turn off the power and release the existing switch from its mounting box. Disconnect the wires from the switch terminals, but leave the earth wire unless a new surface-mounting box is being fitted to take a new 2-way switch.

From the existing switch, run the 3-core cable to the new switch position. The route taken by the cable depends on the relative positions of the two switches. Where both switches are on the ground floor the cable will run under the floorboards in the rooms above and drop down the wall to the switches. Where the switches are in upper rooms or in a bungalow the cable will be run in the roof space. Where one switch is on the first floor and the second switch is on the ground floor the cable will run down the wall from the first switch into the ceiling void and on down the wall to the second switch. Cables run down the wall may be fixed to the surface using mini-trunking or sunk into the plaster.

Having installed the cable, thread the end at the existing switch position into the mounting box together with the existing cable. Where the new cable drops down from the ceiling, it will normally be possible to put it through the grommet alongside the existing cable. Where it runs down the wall to a switch below, you will need to knock out a hole in the bottom of the box.

Fix or refix the box as relevant, leaving the existing and the new cables in the box until the switch is fitted. At the second switch position thread the new cable into the box and mark the position of the box on the wall. If a plastic surface-mounting box, suitable for use with mini-trunking, is being used fit the mini-trunking first and take this into the box before fixing the box to the wall; if the switch is to be flush-mounted, cut the hole for the box and the chase for the cable, fix the box and make good the wall surface.

Trim the cable to about 100mm in length and strip off about 75mm of the outer sheathing. Strip off about 10mm of insulation from the ends of each of the three insulated wires and slip green/yellow PVC sleeving over the bare earth wire. Connect the earth

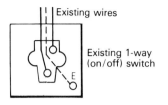
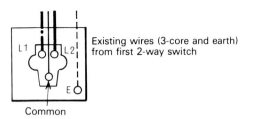
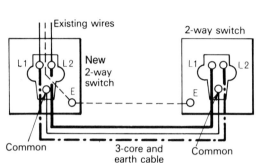
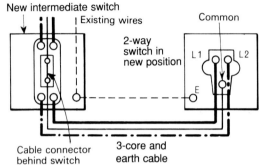

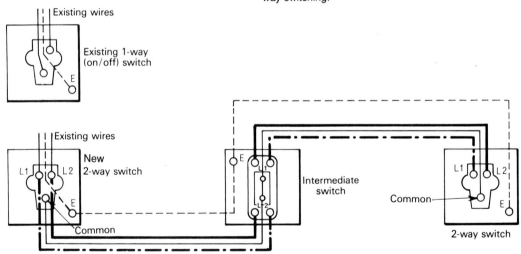

Fig 50 *Above left*: Converting 1-way switching to 2-way switching. *Above right*: Converting 2-way switching to 3-way switching. *Below*: Converting 1-way switching to 3-way switching.

wire to the earth terminal in the base of the box.

The three insulated wires are coloured red, yellow and blue respectively. The 2-way switch has three terminals marked 'Common', L1 and L2 respectively. Connect the red wire to the Common terminal, the yellow wire to terminal L1 and the blue wire to terminal L2. Place the wires in the box and fix the switch to the box using the two screws supplied.

At the first switch position trim the new cable to about 100mm in length, slip green/yellow PVC sleeving over the bare earth wire and connect this wire to the earth terminal in the base of the box. If the existing earth wire is bare enclose this also in green/yellow PVC sleeving and reconnect it to the earth terminal. Strip about 10mm of insulation from the ends of each of the three insulated wires. Connect the red wire to the Common terminal of the switch. Connect the

Lighting controls

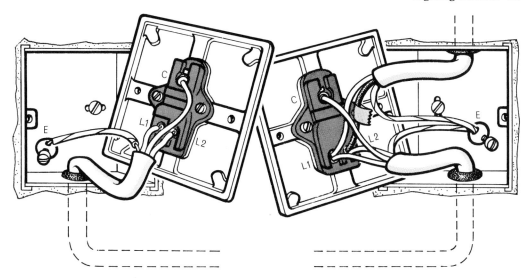

Fig 51 Wire connections at each of the two 2-way switches of a 2-way switching circuit. Each switch is mounted on a flush metal box having an earth terminal for the earth conductors. The two switches are linked by a 3-core and earth PVC-sheathed cable.

yellow wire together with one of the existing insulated wires to terminal L1. Connect the blue wire together with the remaining existing insulated wire to terminal L2 of the switch. (The two existing wires will be red and black respectively but either may be connected to L1 or L2 terminal.)

2-way switch conversion for a bedroom

Where 1-way switching controlling a light in a bedroom is to be converted to 2-way switching it is often more convenient if the second switch is a cord-operated ceiling switch, fixed over the bedhead. The wiring is the same as for two wall-mounted 2-way switches with the exception that at the new switch position the cable passes through a hole pierced in the ceiling instead of running down the wall.

Preferably the ceiling switch should be fixed in line with the bed centre. If there is no convenient joist in that position fix a length of wood batten between and flush with the bottom of the two nearest joists. But first drill a hole in the batten corresponding with the hole in the ceiling to accommodate the cable.

Release the switch from its pattress box. Knock out the thin section of plastic from the base of the box and thread in the cable. Fix the box to the ceiling making sure that the screws are driven into the wood batten. Strip off about 100mm from the outer sheathing of the cable. Slip green/yellow PVC sleeving over the end of the bare earth wire and connect this wire to the earth terminal in the base of the pattress box. Strip about 10mm of insulation from the ends of each of the three insulated wires. Connect the red wire to the Common terminal of the 2-way ceiling switch. Connect the yellow wire to terminal L1 and the blue wire to terminal L2 of the switch. Place the wires neatly in the box and fix the switch, using the two screws earlier removed from the box.

The connections at the 2-way wall switch are the same as already described.

New 2-way switching circuits

When installing an entirely new light which is to be controlled by 2-way switching, the wiring is the same as for a conversion from 1-way switching to 2-way. A twin and earth PVC-sheathed cable is run to the first of the two switch positions as though for 1-way switching and from this first switch a length of 3-core and earth PVC-sheathed cable is run to the second switch position.

The materials required for the light and the 2-way switching circuit are: loop-in ceiling rose, flexible cord and lampholder (or other light fitting if not a plain pendant); two

96 Lighting controls

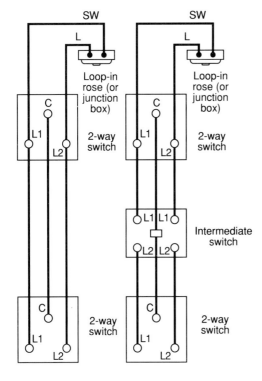

Fig 52 *Left*: A 2-way switching circuit. *Right*: An intermediate switching circuit. The earth conductor and mains supply wires are omitted for clarity.

light need to be switched from more than three separate positions.

The intermediate switch, which has four terminals (two L1 and two L2 terminals), is inserted into the two strapping wires which link the L1 and L2 terminals of the two 2-way switches (see Fig 52). When using 3-core and earth PVC-sheathed cable the yellow and the blue wires usually link the L1 and L2 terminals of the two 2-way switches; these are the strapping wires. The red wire linking the Common terminal of each switch is not a strapping wire and is not connected to the intermediate switch but is joined in a cable connector in the mounting box.

The materials required for the installation of an intermediate switching circuit (in addition to the light fitting) are: two 2-way 6A rocker plateswitches; one intermediate rocker plateswitch; three switch mounting boxes, either surface or flush fixing; a length of 1.0mm^2 2-core and earth flat PVC-sheathed cable; a length of 1.0mm^2 3-core and earth flat PVC-sheathed cable; wall plugs, cable clips and woodscrews; one cable connector (terminal strip).

Wiring the switching circuit

Select the positions for the three switches and plan the route for the cable running from the light to the first switch and the cable linking the three switches.

From the loop-in ceiling rose (or junction box depending on the wiring), run a length of 2-core and earth PVC-sheathed cable to the first switch position. From this switch position to the next run a length of the 3-core and earth PVC-sheathed cable and from this second switch position run a length of 3-core and earth PVC-sheathed cable to the third switch.

Prepare each of the three mounting boxes for the cable entry allowing for two cables at the first and second switch. Cut the holes in the plaster if any of the boxes are to be flush mounted, plus chases to run the cable in the wall.

Thread the two sheathed cables into the first switch box and fix the box to the wall; thread the two sheathed cables into the second box and fix this box to the wall; thread the single sheathed cable into the third box and fix this box to the wall. Make good the plaster before wiring.

2-way 6A rocker plateswitches; two switch mounting boxes (surface or flush fixing); a length of 1.0mm^2 2-core and earth flat PVC-sheathed cable; a length of 1.0mm^2 3-core and earth flat PVC-sheathed cable; a short length of green/yellow PVC sleeving; wall plugs, woodscrews and plastic cable clips. If using the junction-box wiring method, you will also need a 20A junction box.

3-way switching circuits for lighting

There are some situations where it would be convenient to switch a light on and off from *three* positions instead of from two positions by means of a 2-way switching circuit (see Figs 50 and 52). For 3-way switching an intermediate switching circuit is installed. This system is actually a 2-way switching circuit with a special switch (termed an intermediate switch) inserted in the cable linking the two 2-way switches. By inserting additional intermediate switches into a circuit a light can be controlled from as many additional positions as there are intermediate switches. However, rarely in the home does a

Connecting the switches

At the first switch position there are two sheathed cables: one 2-core and one 3-core both with an earth wire. Trim these to a length of about 100mm. Strip off about 75mm of outer sheathing from the two cables. Strip about 10mm of insulation from the five insulated wires. Enclose the two bare earth wires in green/yellow PVC sleeving and connect both to the earth terminal in the base of the box.

Take one of the two 2-way switches. Connect the red wire of the 3-core cable to the Common terminal of the switch. Connect the red wire of the 2-core cable and the yellow wire of the 3-core cable to terminal L1 of the switch. Connect the black wire and the blue wire to the L2 terminal of the switch. Place the wires neatly in the box and fix the switch to the box using the two screws supplied.

At the second switch position there are two 3-core and earth PVC-sheathed cables (see Fig 50). Trim each cable to length and strip off about 75mm of outer sheathing from each. Slip green/yellow PVC sleeving over the bare earth wires and connect them to the earth terminal in the box. Strip off about 10mm of insulation from the ends of each of the six insulated wires. Take the insulated cable connector and to it connect the two red wires. Now take the intermediate switch. Connect the yellow and blue wires of one of the two cables to the top two terminals of the switch. Connect the remaining yellow and blue wires to the bottom two terminals of the switch. Place the wires neatly in the box and position the cable connector so it will not foul the switch. Fix the switch to its box using the two screws supplied.

At the third switch position there is only one 3-core PVC-sheathed cable. Trim this to length, strip off 75mm of outer sheathing and about 10mm of insulation from the three insulated wires. Slip green/yellow PVC sleeving over the earth wire and connect this wire to the earth terminal in the base of the box. Connect the red wire to the Common terminal of the remaining 2-way switch, connect the yellow wire to terminal L1 and the blue wire to terminal L2 of the 2-way switch. Place the wires in the box and fix the switch to the box using the screws supplied with the fitting.

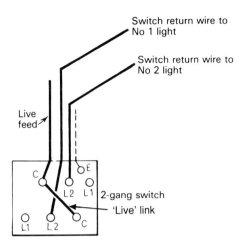

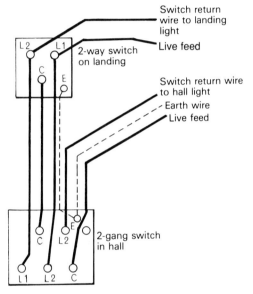

Fig 53 *Top*: A 2-gang switch with both switches wired as 1-way for switching two lights independently from the one switch position. *Bottom*: One 2-way switch of the 2-gang switch is wired as a 2-way for the landing light in conjunction with the single 2-way switch on the landing. The other 2-way switch is wired for 1-way switching of the hall light.

Installing a twin (2-gang) lighting switch

Where two lights are to be switched independently but from the one switching position a 2-gang switch is used. The two switches are mounted on a single 'square' 1-gang faceplate and the complete unit is mounted on a 1-gang switch box. Both switches are of the 2-way version though either or both can also be used as 1-way switches. The wiring to the switch unit depends on whether the switches are to be used for 1-way or 2-way switching.

98 Lighting controls

Fig 53 shows the wiring at a 2-gang switch for two 1-way switches which is best wired using junction box wiring. A 6-terminal junction box is used with terminals for each of the two switch return wires (see Fig 54). The red wire of the cable is connected to the Common terminal of one switch and from this terminal a short length of red insulated wire is run to the Common terminal of the other switch, this being the live feed wire linking the two switches. The yellow wire which is the switch return wire of one light is connected to the L2 terminal of one switch. The blue wire which is the switch return wire of the other light is connected to the L2 terminal of the other switch. The L1 terminal of both switches remains blank. When fixing the switch assembly to its mounting box make sure that the switch is the right way up; see 'Top' at the back of the faceplate.

When one switch of a 2-gang unit is used as a 1-way switch, the other as a 2-way switch (as in the hall or the hall and landing lights) each switch is treated independently. A 2-core and earth PVC-sheathed cable runs to the 1-way switch, and a 3-core and earth PVC-sheathed cable to the other switch, there being no electrical connection between the two switches. See Fig 53.

Where both switches of a 2-gang unit are used as 2-way switches these too are treated as separate circuits but the wiring connection at each switch will depend upon whether it is the first or second switch of the 2-way circuit.

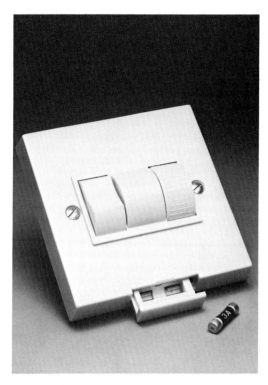

Photo 40 Combined dimmer and rocker switch fitted with a fuse to prevent damage to dimmer from current surges when a bulb fails (*MK*).

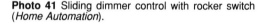

Photo 41 Sliding dimmer control with rocker switch (*Home Automation*).

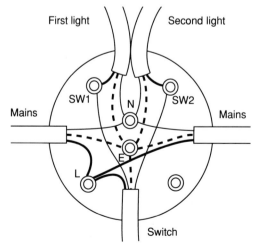

Fig 54 Connections at 6-terminal junction box for two lights switched from a 2-gang switch.

Lighting controls 99

Dimmer switches

The dimmer switch is an electronic switch which when fitted into a lighting circuit enables the intensity of the light to be varied from full intensity down to zero and OFF. It can be used only with tungsten lights – ie *not* fluorescent lights.

The home dimmer switch is mounted on a square plate of the same dimensions as the modern plateswitch. It can be mounted on the standard 1-gang box which can be either a moulded plastic surface box or a metal box sunk into the wall.

The standard dimmer with a rotary control is designed to replace a 1-gang plateswitch with 1-way operation. 2-gang versions are also available (plus 3-gang and 4-gang to fit in double mounting boxes) for controlling more than one light from the same position.

For 2-way (or 1-way) operation, push-button dimmers leave the intensity of the light at the set position, an effect also achieved with slide dimmers and dimmers with a separate rocker switch. Touch-sensitive dimmers (and touch switches) are available in 1-gang or 2-gang versions for 1-way or 2-way operation.

Photo 42 Touch-sensitive dimmer switch (*Super-switch*).

To install a dimmer, the existing light switch is removed with the power off and the existing wires connected to the dimmer switch terminals, following the manufacturer's instructions. For dimming a table lamp, a table lamp dimmer can be fitted in the flex to the lamp.

Photo 43 Range of rotary dimmer switches (*Super-switch*).

100 Lighting controls

Photo 44 Table lamp dimmer (*Home Automation*).

Photo 45a Remote-control dimmer (*Home Automation*).

Photo 45b Hand-held controller for remote-control dimmer (*Home Automation*).

Time delay switches

A time delay switch will reduce the light level slowly to a selectable lower level (or turn it off completely) after a pre-set time – typically 2 to 15 minutes.

Remote-control switches

Lights in the home can be dimmed or turned on and off using a battery-operated hand-held infra-red transmitter. This is pointed at the wall switch (or at a receiver on the ceiling) which then operates the light. The hand-held controller, which has a range of 10–15m, needs to be in the line of sight of the receiver, but will operate through (clear) glass.

The wall-mounted switch can also be operated by hand (touch operation) and remote-control switches can be used to operate appliances other than lights up to a maximum load of 5A (1200W).

For use outside, splashproof remote-control switches are available.

Photo 45c Exterior remote-control switch (*Home Automation*).

Photo 45d Remote-control light switch and hand-held controller (*Superswitch*).

Security light switches

One way of making your home more secure is to give the impression that you are in when you are out. A security light switch can do this by bringing on lights automatically.

A simple security light switch has a photocell and will bring the light to which it is connected on at dusk and turn it off again at dawn. A *programmable* security light switch can be set to turn the light on at dusk and then off again after a pre-set time typically ranging from 1 to 8 hours. With more sophisticated models, six or more on/off periods can be programmed or the switch can be set to give *random* switching. *Copy* switches 'remember' the switching pattern over a 24-hour period and will then repeat this the following day.

A security light switch simply replaces an existing light switch and can be used in both 1-way and 2-way switching circuits. Most are 1-gang switches, but 2-gang versions are available with one of the two switches being an ordinary switch. All security switches can be overridden manually if required.

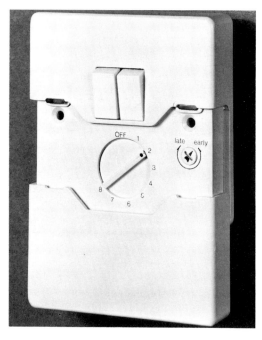

Photo 46a Programmable security light switch with built-in photocell (*Smiths Industries*).

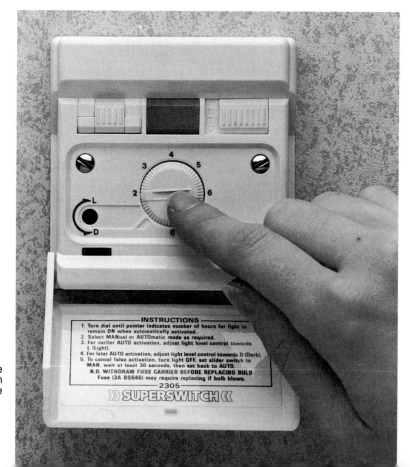

Photo 46b Programmable security light switch with built-in photocell and fuse (*Superswitch*).

Timer switches

A plug-in timer switch can be used for security lighting – by turning a table lamp on and off, say – but can also be used for operating other electrical appliances, such as a tape recorder.

Most timer switches are plug-in – that is, they fit into a 13A socket outlet and have their own socket to take the plug of the lamp or appliance – but you can also get timers which you wire in permanently (for controlling an immersion heater, for example). Both 24-hour and 7-day versions are available and there is a choice between electro-mechanical operation (where you move small tappets to set the ON and OFF times) and electronic operation (where push buttons are used to set the times and very accurate setting can be achieved). Some electronic timers offer random switching and all have a built-in battery which preserves the programme if the time switch is unplugged.

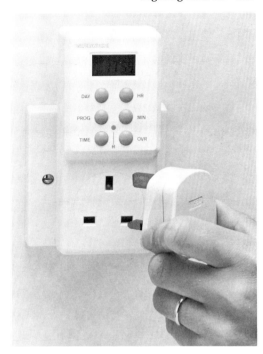

Photo 47 Electronic 7-day plug-in timer switch (*Superswitch*).

Photo 48 Electro-mechanical 24-hour plug-in timer switch (*Superswitch*).

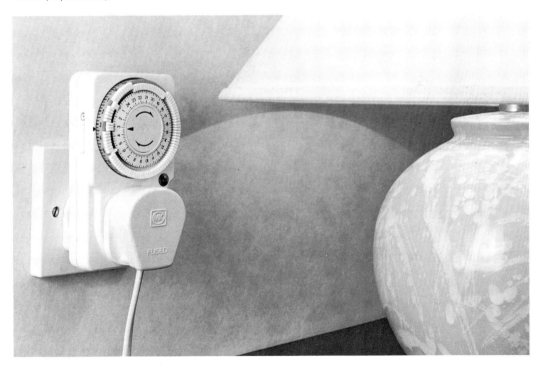

7: Socket outlets

The standard socket outlet installed in British homes is the 13A 3-pin socket and fused plug, with its 'square' pins. Although the 13A fused plug and socket has been in use for many years, a few homes may still have some of the earlier round-pin plugs and socket outlets which had current ratings of 2A, 5A and 15A.

Any socket outlets like this should be replaced with modern 13A sockets, preferably on a new rewired circuit. New 2A round-pin sockets can be used for table lamps – see page 175 for details.

The 13A socket outlet

A 13A socket outlet is made in various versions, that is: switched, non-switched, switched with neon indicator. All versions are made in both single and double socket units. Most single socket outlets are of the 'square' plate type and fit one-gang moulded plastic surface boxes and metal flush boxes. Double socket outlets fit the equivalent surface and flush 2-gang boxes.

Most makes and models of the 13A 'square' socket outlet fit standard boxes having a 25mm depth but some require the 35mm depth boxes to provide extra wiring space when needed.

Most socket outlets are white plastic, but you can get coloured plastic, brass, aluminium, steel and hardwood finishes.

The 13A fused plug was designed initially for use with the ring circuit but it is now also used with multi-outlet radial circuits, not wired in the form of a ring.

Replacing a damaged 13A socket outlet

To replace a cracked, broken or faulty (burnt-out) 13A socket outlet is usually a simple job taking only a few minutes.

First buy a new socket outlet which matches the existing socket outlet in shape, style and colour. If the new socket does not match the old socket and it is surface-mounted buy a new box to match the new socket, preferably the type specifically designed for use with mini-trunking. If mounted on a flush box, it will not be necessary to match the old socket since all makes and styles will fit over the metal box sunk into the wall.

Photo 49 Unswitched 13A socket outlet (*MK*).

Photo 50 Switched 13A socket outlet (*MK*).

Socket outlets 105

Photo 51 Single and double switched socket outlets, with and without neon indicators (*MK*).

Photo 52 Co-ordinated range of electrical accessories, including plastic and metal-finish socket outlets (*PowerBreaker*).

Socket outlets

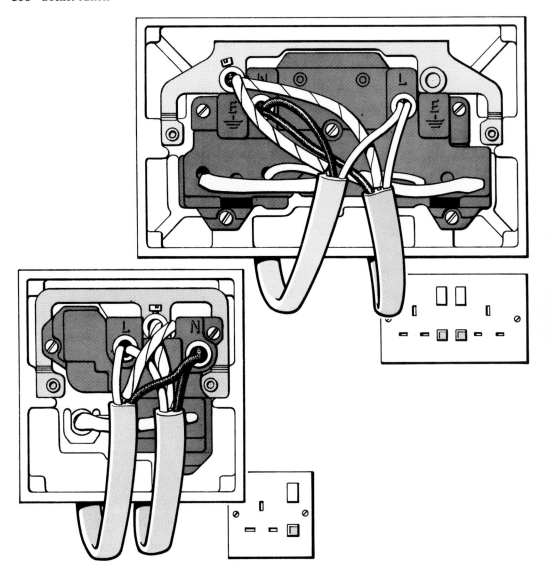

Fig 55 *Above*: Connections of ring circuit cables at a double socket outlet with neon. *Left*: Connections of ring circuit cables at a single socket outlet with neon.

Turn off the power and remove the circuit fuse, which is most likely a ring circuit fuse. Plug a portable lamp (table lamp) into the socket to check that the socket is 'dead'. If it is a faulty socket which is 'dead' anyway and cannot be tested, remove all the circuit fuses as a precaution, in case someone should turn on the main switch before the job is completed.

Remove the two fixing screws and release the socket from its box. Disconnect the wires from the terminals – usually there are two red, two black and two earth wires. Loosen the three terminals of the new socket outlet. Connect the red wires to the 'L' terminal, the black wires to the 'N' terminal and the earth wires to the 'E' terminal (see Fig 55).

The earth wires should be enclosed in PVC sleeving, usually green. If no sleeving is present, slip green/yellow PVC sleeving (which has replaced the green) over the bare earth wires before re-connecting them to the earth terminal.

Where the existing box has been retained, fix the socket to the box using the old screws. If a new box, use the screws which came attached to the new socket outlet. The old screws are likely to be 4BA Imperial screws, the new ones M3.5 metric screws.

To replace a moulded plastic box with a new box, remove the old box by releasing the two wood screws. Knock out a section of the thin plastic at the base of the box, or from the edge if surface wiring (unless a mini-trunking box is being used). Thread in the cables and fix the box to the wall using new screws. Connect and fix the socket outlet as described above. Turn on the power and test the new socket outlet.

Converting a surface socket outlet to a flush version

The standard 'square plate' single 13A socket outlet and its double equivalent are designed for mounting on either a flush box or a surface box as required. Flush mounting (see Fig 56) is usually preferred as it is neater and does not project so far from the wall. A flush-mounted socket, and its plug when fitted, is less likely to be damaged by furniture and cleaning tools.

Provided the existing surface-mounted socket outlet is of the 'square plate' type, the only materials required for converting it to flush mounting are: a flush metal box; two PVC grommets; two screws and wall plugs; and possibly a short length of green/yellow PVC sleeving.

First turn off the power, and remove the circuit fuse. Release the fixing screws and remove the socket outlet from its box. Disconnect the wires from the socket terminals and remove the box from the wall. Hold the flush box in position and mark its desired position on the wall by running a pencil around the box and draw two parallel lines downwards from the centre of the box for the cable chase. Cut along all the lines with a sharp trimming knife, peel the square and strip of wallpaper off the wall and, with an electrician's bolster (or a sharp cold chisel) and a club hammer, chop out the plaster down to the masonry or brickwork and clean out all dust and debris. Try the box in its hole and, if it projects, chop out some of the masonry until the box sits neatly in the hole firmly against the masonry. Lay the cable in its wall chase (if necessary, make a slot behind the skirting board to take the cable before you do this). Knock out two cable

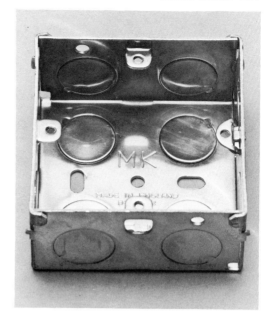

Photo 53 Standard flush fitting 1-gang box for socket outlets and other 1-gang wiring accessories (*MK*).

Photo 54 Standard surface fitting 1-gang box for socket outlets and other 1-gang wiring accessories (*MK*).

entry blanks on one edge of the box, (usually the bottom edge) and fix a PVC grommet into each knockout hole. Place the box in the hole, mark two fixing holes, drill and plug the holes. Thread in the cables and fix the box using No 8 wood screws.

If the box is not level or 'rocks' in the hole, cut the hole slightly deeper, then wet the brickwork and insert some plaster filler. Press in the box until it fits square and flush. After about ten minutes remove the box to

108 Socket outlets

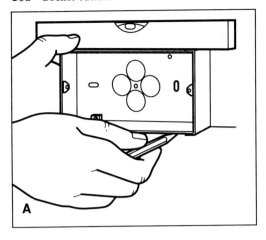
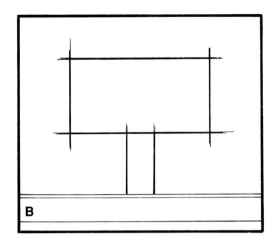
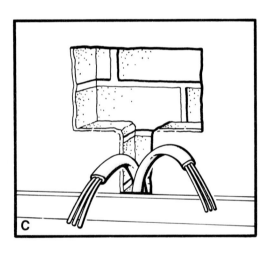
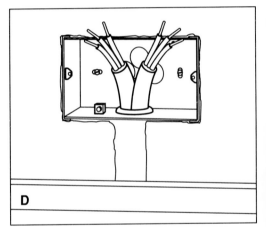
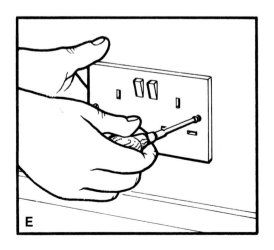

Fig 56 The steps in fitting a flush-mounted socket outlet: **A** Mark the wall for the hole and the chase for the cable; **B** Cut along the marked lines with a trimming knife; **C** Cut out the hole and chase with a cold chisel or bolster; **D** Thread the cable through the grommet (or grommets) in the box, fit the box to the wall and make the wall good with filler; **E** Wire up the socket outlet and secure it to the box.

allow the filler to harden. When hard, drill two holes in the wall, selecting one or both slotted holes in the base of the box so the box can be adjusted if necessary to get it level. Plug the holes and refix the box, after first threading the cables through the grommet. Make good the gap around the box and over the cable chase with filler and allow the filler to set before wiring up the socket outlet. For fixing flush-mounted socket outlets to partition walls, see pages 85 and 86.

Connect the wires to the socket terminals: red wires to the 'L' terminal, black wires to the 'N' terminal and the earth wires to the 'E' terminal of the socket outlet. If the earth wires are bare, first enclose them in green/yellow PVC sleeving.

If the metal box has an earth terminal, the earth conductors of PVC-sheathed cable are *not* connected to this terminal – they must be connected to the 'E' terminal of the socket outlet.

Place the wires in the box and fix the socket outlet using the two M3.5 metric screws supplied (not those used with the surface box as they may be 4BA screws).

Photo 55 Rear view of 13A socket outlet showing terminals (*MK*).

Converting a single socket outlet to a double

Converting single 13A socket outlets to doubles increases the number of socket outlets without having to run additional wiring. With a few possible exceptions every single 13A socket outlet on a ring circuit may be converted to a double. The exception is where two 13A socket outlets have been supplied from a ring circuit spur. This used to be allowed by the wiring regulations, but the current permitted maximum allowed is *one* single or double socket outlet. A socket outlet connected to a spur has only one cable instead of two, but *the first* socket outlet on a 2-socket spur has two cables so without first testing the wiring (see Fig 61), it is not possible to tell whether the socket is a ring socket or a spur type.

Converting a surface single socket to a surface double socket

The materials required for converting a surface single socket outlet to a surface double socket outlet are: one double socket outlet (preferably switched); mini-trunking; one plastic double mounting box; some green/yellow PVC sleeving (may be required); wall plugs and two wood screws.

First turn off the power. Remove the fixing screws of the single socket and disconnect the wires from the socket terminals. Unscrew the box from the wall by removing the wood screws. (If the screws are rusted-in break the box and use pliers to remove the screws.) Knock out a section of thin plastic (unless a mini-trunking box is used). Fit mini-trunking to take the cable and drill and plug holes in the wall to take the fixing screws for the box.

Thread in the cables and fix the box to the wall. Check that the box is level. If not and the slotted screw fixing holes have been used, level the box using a small spirit level if necessary. Connect the wires to the socket terminals: reds to the 'L' terminal, blacks to the 'N' terminal and the earth wires to the 'E' terminal. If necessary, slip green/yellow PVC sleeving over the earth wires before connecting them. Place the wires in the box and fix the socket to the box using the two screws supplied with the socket outlet.

110 Socket outlets

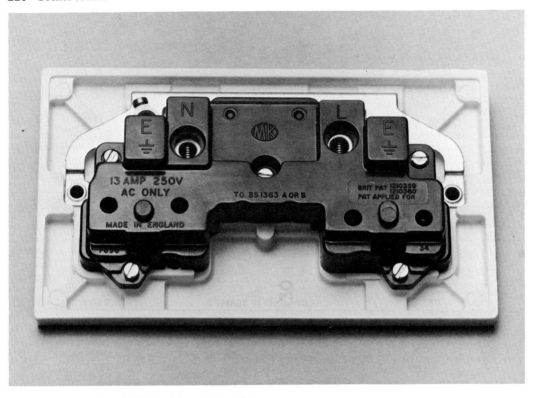

Photo 56 Rear view of 13A double socket outlet showing terminals (*MK*).

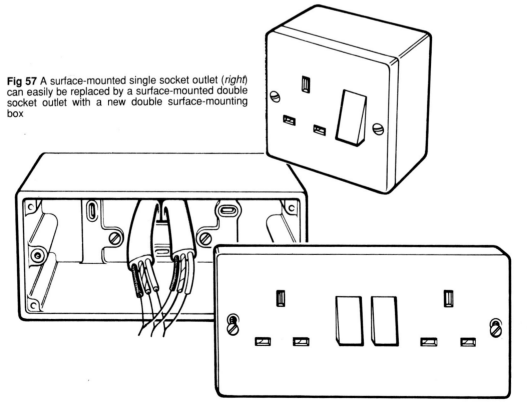

Fig 57 A surface-mounted single socket outlet (*right*) can easily be replaced by a surface-mounted double socket outlet with a new double surface-mounting box

Socket outlets

Photo 57 A double and a single 13A socket outlet mounted on surface boxes (*MK*).

Converting a flush single socket to a flush double socket

To convert a single flush socket outlet to a double flush type requires one double socket outlet; a 2-gang standard metal mounting box; two PVC grommets; fixing screws; wall plugs and plaster filler.

First turn off the power. Remove the single socket from its box by releasing the two fixing screws. Disconnect the wires from the socket terminals. Remove the wood screws fixing the existing metal box. Try to ease out the box using an old screwdriver if necessary to prise it from the brickwork. If the box is cemented in chop out the cement using a sharp cold chisel but take care not to damage the cables running into the box.

When the old box has been removed hold the 2-gang box over the hole and mark along the edges with a pencil to indicate by how much the hole is to be widened. Using a short sharp cold chisel chop out the plaster and masonry within the pencil lines until the new box fits flush with the wall. Remove one (or two) knockouts and fit a PVC grommet (or grommets) into the knock out hole(s).

Place the box in the hole, mark and drill holes for wallplugs, thread the cable through the grommet(s) and secure the box with No 8 screws. Make good the hole around the box before wiring. Enclose any bare earth wire in green/yellow PVC sleeving. Connect the red wires to the socket 'L' terminal, the black wires to terminal 'N' and the sleeved earth wires to the 'E' terminal of the socket outlet. Place the wires in the box and fix the socket outlet to the box using the two screws supplied.

Converting a flush single socket to a surface double socket

To save chopping out a flush single box and sinking a 2-gang box in its place (which in many situations is not possible) a single flush socket outlet can be replaced by a surface double. This is a simple job taking very little

112 Socket outlets

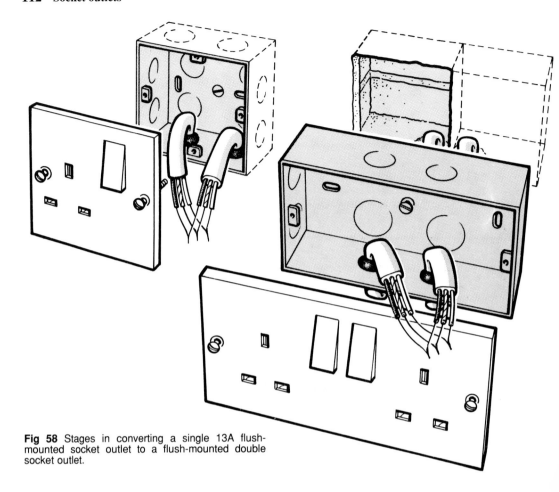

Fig 58 Stages in converting a single 13A flush-mounted socket outlet to a flush-mounted double socket outlet.

time. The materials required are: one double socket outlet and one 2-gang moulded plastic surface box or single-to-twin converter.

First turn off the power and remove the socket outlet from its one-gang metal box. Disconnect the wires from the socket. Knock out the thin section of plastic in the base of the surface box.

Thread in the cables and hold the box over the flush metal box with the centre of the double box in line with the metal box where it will be seen that four fixing holes in the base of the plastic box coincide with the four screw lugs of the metal box. A single-to-twin converter has a metal plate which lines up with the two side lugs. Use the two screws taken from the old single socket, fix the plastic box or converter to the metal box taking care not to pierce the cables. Connect the wires to the new socket and fix the socket to the box using the screws supplied with the socket outlet.

Replacing a single socket by two singles

After converting a number of single socket outlets to doubles one is left with a quantity of single socket outlets still in a good condition. These can be utilised by mounting two side-by-side on what are termed dual boxes.

A dual mounting box, made in moulded plastic and in metal for surface and flush mounting respectively, is slightly wider than the 2-gang box used for double socket outlets. The box also has two extra screw lugs in the centre to take one screw from each of the single sockets. These boxes are also used for a variety of other wiring accessories where two are required in the one position.

The materials required for the conversion are: one dual box, surface or flush mounting;

Photo 58 A 2-gang surface box for a double socket outlet (*MK*).

Fig 59 Stages in converting a single flush-mounted 13A socket outlet to a surface-mounted double socket outlet.

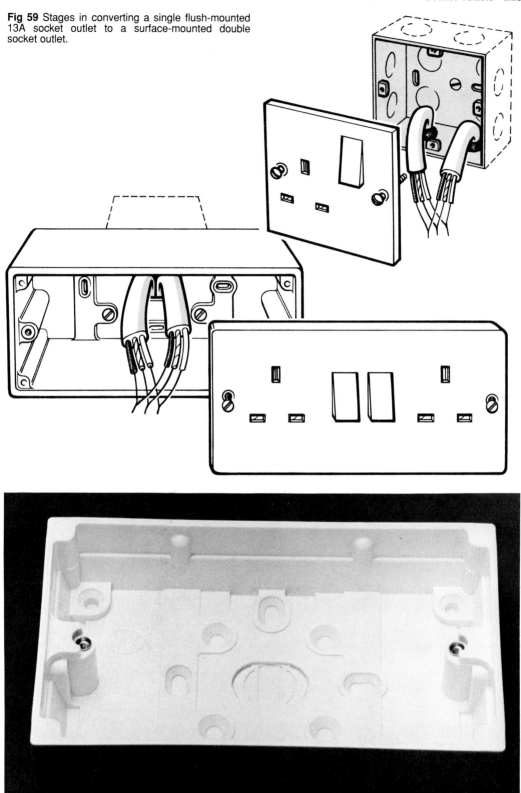

114 Socket outlets

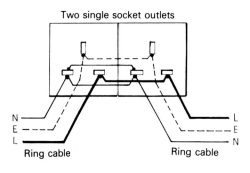

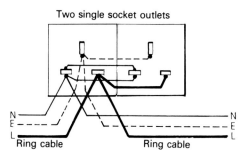

Fig 60 Alternative methods of connecting two single 13A socket outlets to a ring circuit where they replace one single socket outlet. In both cases, the two socket outlets are linked by a short length of 2-core and earth cable; in the first, preferred, method (*top*), the two socket outlets are part of the ring circuit, in the second method, the right-hand socket outlet is a spur.

Photo 59 A dual surface box for two 1-gang sockets or two other wiring accessories (*Crabtree*).

four M3.5 fixing screws; about 250mm of 2.5mm^2 2-core and earth PVC-sheathed cable; a length of green/yellow PVC sleeving; wall plugs and wood screws.

As two single socket outlets on one box have not the smooth finish of a double socket outlet and are therefore less attractive they should be fitted in the less important rooms or in situations where they are hidden from view. If you choose a surface box, you will save the work of sinking a flush-mounting box into the wall. The dual box can replace a single surface box or it can be fixed over a 1-gang metal flush box as it has fixing holes which coincide with the screw lugs of a 1-gang metal box.

First turn off the power and remove the single socket outlet from its box. If a surface box remove this also; if a flush box retain it in position.

Knock out the section of thin plastic from the base of the dual box. Thread in the cables and fix the box, using wood screws in drilled and plugged holes when fixing direct to the wall and 4BA screws when fixing to an existing 1-gang flush box.

Of the two existing cables connect the wires of one cable to the terminals of one socket outlet: red to 'L' terminal, black to 'N'

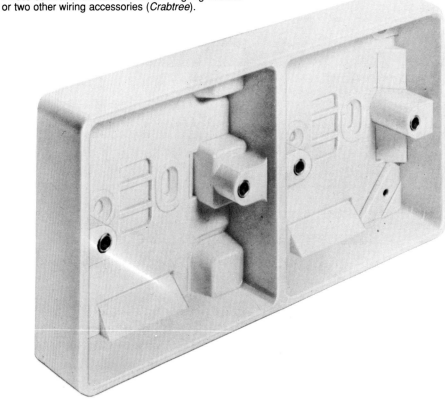

Socket outlets

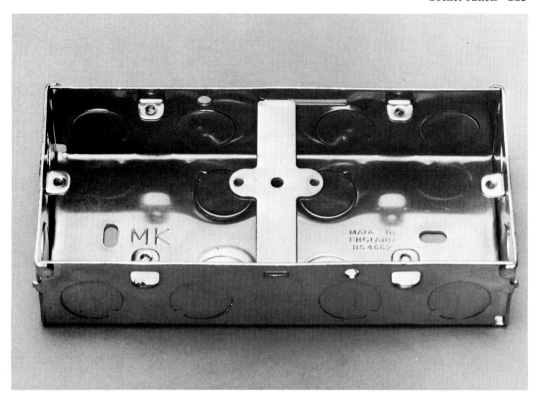

Photo 60 A dual flush box for two 1-gang socket outlets or two other wiring accessories (*MK*).

terminal, earth wire to 'E' terminal. Connect the wires of the other cable to the corresponding terminals of the other socket outlet.

Strip off the outer sheathing from the new length of 2.5mm² 2-core and earth PVC-sheathed cable. Connect one end of the red wire to the 'L' terminal of one socket outlet, the other end to the 'L' terminal of the other socket outlet. Similarly connect a black wire to the 'N' terminal of each socket outlet. Slip green/yellow PVC sleeving over the bare earth wires and connect them to the 'E' terminals of the respective socket outlets.

Connecting the two socket outlets in this manner places both in the ring circuit, making them ring sockets rather than one being a spur socket outlet (see Fig 60). Place the wires in the box and fix the sockets to the dual box using the new M3.5 metric screws.

Moving a 13A socket outlet

When an existing socket outlet becomes obstructed by furniture or is in the wrong position in relation to the appliance it serves, or for any other reason, it can be moved to a new position. There are two principal methods of moving a socket outlet.

First method

The materials required are: one 3-way 20A cable connector; one cover plate for the old socket outlet box; a length of 2.5mm² 2-core and earth flat PVC-sheathed cable; one socket outlet mounting box; cable clips, wood screws and two M3.5 metric screws.

The method involves removing the existing socket outlet but using the existing mounting box, fitted with a cover plate, as a junction box. Cable is run from the existing socket outlet to the new socket outlet position.

Lift a floorboard in close proximity to the existing socket outlet; lift a floorboard below the new socket outlet position and raise any other floorboards as necessary to facilitate running the cable from one position to the other.

116 Socket outlets

At the new socket position chop out the plaster behind the skirting board to accommodate the new cable. Do likewise at the existing position taking care not to damage the existing cables. Drill the joists as necessary and prepare the way for the new cable.

Run the cable along its route; pass one end up behind the skirting board at the new position and pass the other end up behind the skirting board at the old position. Prepare the box for the new cable, thread in the cable and fix the box to the wall, either on the surface or sunk into the wall depending upon the type of box chosen. Where the socket outlet is to be fixed on the same wall as its original position, the cable can be run behind the skirting board or run in mini-trunking so saving the trouble and difficulties of lifting floor covering or raising floorboards. Mini-trunking systems also have corner pieces enabling the cable to be run around a corner as well as special mounting boxes.

Turn off the power, and remove the socket outlet from its original box but leave the box in position to be used as the junction box. Thread in the end of the new cable and prepare the end for connection in the cable connector. Slip green/yellow PVC sleeving over each of the earth conductors. Connect these earth wires to the centre terminal of the 3-way cable connector. Connect the red wires to one of the outer terminals of the connector. Connect the black wires to the remaining terminal. Place the wires in the box and fit the cover plate using the original screws.

At the new socket position prepare the end of the cable. Connect the red wire to the 'L' terminal of the socket outlet, connect the black wire to the 'N' terminal and the sleeved earth conductor to the 'E' terminal of the socket outlet. Place the wires in the box and fix the socket outlet using the two new M3.5 metric screws.

Alternative method

Instead of retaining the existing socket outlet mounting box for use as a junction box, the box as well as the socket outlet is removed and the two ring circuit cables are wired to a 30A 3-terminal junction box where they pass up behind the skirting board. With the power turned off, cut the cables here and also at the old socket outlet. Fit the junction box to a length of timber secured between joists.

Connect the six wires to the terminals of the junction box together with the three wires of the new 2.5mm² cable. The three earth conductors are connected to the centre terminal and the red and black wires are connected to the outer terminals respectively.

Screw on the cover of the junction box and check that the sheathing of all three cables terminates within the box so that no unsheathed wires are exposed.

The remainder of the work is the same as for the previous method. With either method make good any breaks in the plaster using plaster filler.

Adding an extra socket outlet or fused connection unit

When an additional socket outlet or fused connection unit is required, this can be added on a spur from the existing ring circuit. The cable to use for the spur is the same as the ring circuit – that is 2.5mm² 2-core and earth PVC-sheathed.

There may be as many non-fused spurs connected to the ring as there are socket outlets on the ring itself and each spur can have one single or double socket outlet or one fused connection unit.

A ring circuit, when installed initially, rarely has many spurs, if any, so there is plenty of scope for extra socket outlets to be wired on spurs. It is first necessary to test the wires at an existing socket to determine whether it is a ring or a spur socket.

To make the test, turn the power off and check the number of cables behind the socket outlet. If only one, the socket is already a spur which cannot be further extended; if two, disconnect the two red wires and connect them to the two probes of a continuity tester (or torch bulb and battery). If the tester or torch bulb lights, the socket outlet is on the ring circuit; if it does not, the socket outlet is the first of two on an old spur and cannot be used for a new one.

The connection of each of the spur cables may be made either at the terminals of a socket outlet connected to the ring cable, or at a junction box inserted into the ring cable. The choice depends on which method would be the easier. Running a cable into an existing socket outlet may mean chopping out a flush box and disturbing the cables

Socket outlets 117

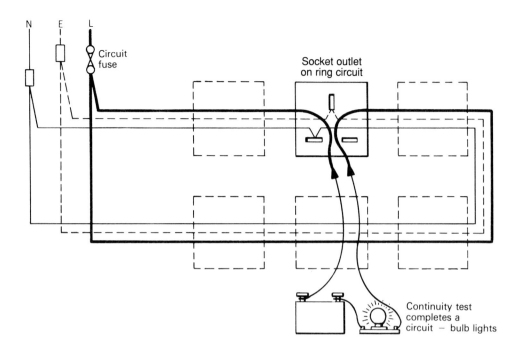

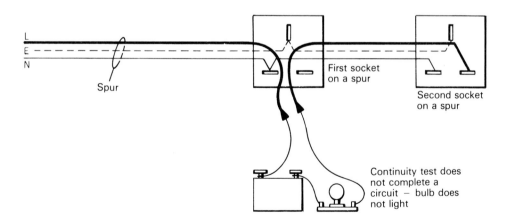

Fig 61 Using a torch bulb and battery to check whether a 13A socket outlet supplied by two cables is a ring socket to which a new spur can be connected or whether it is the first of two spur sockets to which no further spurs may be connected. An easier instrument to use would be a continuity tester or a multimeter.

118 Socket outlets

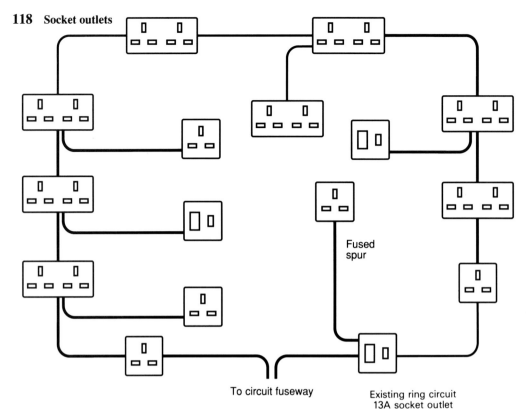

To circuit fuseway

Fig 62 Non-fused spurs connected to a ring circuit supplying extra socket outlets and fused connection units (for fixed appliances) plus a fused spur (see page 136).

Fig 63 Connecting a spur to the ring circuit cable to feed a single or double 13A socket or a fused connection unit. *Top*: Spur cable joined to existing socket outlet. *Bottom*: Spur cable joined to 30A junction box.

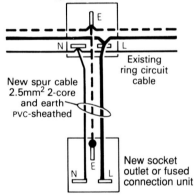

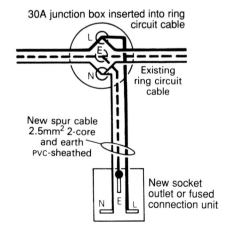

buried in the wall. Where the cable supplying the existing socket outlet is run on the surface and the socket outlet is mounted on a surface box, running a new cable into the box would not be difficult. Equally, if a socket outlet is required the other side of an internal wall from an existing flush socket, the spur cable can be run out of the *back* of the existing flush-mounting box.

The alternative, which is to insert a junction box into the ring cable, is not normally difficult provided the floorboards can be raised to get at the cable and fix a junction box.

The materials required for adding a 13A socket outlet to the ring circuit are: a length of 2.5mm² 2-core and earth flat PVC-sheathed cable; green/yellow PVC sleeving; one 13A single or double socket outlet (or fused connection unit); mounting box, single or

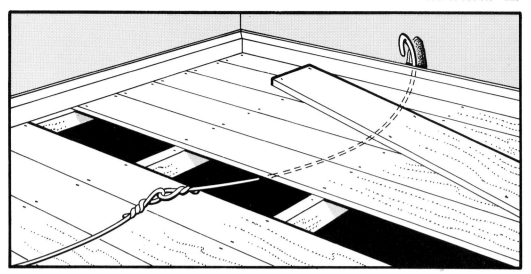

Fig 64 'Fishing' a wire under floorboards to save lifting a number of floorboards in order to run a cable behind a skirting board up to a socket outlet. The circuit cable is attached to the draw wire.

double, flush or surface; cable clips, wood screws, wall plugs, one PVC grommet if a flush (metal) mounting box is used; also, if the junction-box method is used, a 30A 3-terminal plastic junction box and a piece of 100mm × 25mm timber.

Installing the cable

Having decided where the connection of the new cable is to be made plan the route for the cable running to the new socket outlet. Where there is access beneath the floorboards, ring circuit cables are normally run under the floor of the rooms where the socket outlets are installed. On a ground floor having a suspended floor the cables can be run under the joists and rest on the ground. Where the floor is solid or is tiled the cables are usually run in the ceiling void under the floorboards in the room above, or in the loft of a bungalow, and 'dropped' down the walls to the socket outlets. On upper floors, such as the first floor of a conventional house, the cables are run under the floorboards in the ceiling void and passed up behind skirting boards to the socket outlets.

When adding socket outlets the cables will follow the usual route, but where the floor is solid, the cable can be run in mini-trunking mounted above the skirting board, in dado-height trunking, in skirting trunking which replaces the existing skirting board or behind skirting board covers – see pages 36 and 37 for more details.

Looping the spur cable into a socket outlet

Turn off the power and remove the socket outlet in which the new cable is to be looped. If a surface box, remove this also and enlarge the cable entry hole to accommodate the new cable and fit mini-trunking.

Run the cable up behind the skirting board from the void below. Thread the end into the box and refix it. Prepare the end of the cable and connect the new wires together with the existing wires into the respective terminals of the socket outlet. Refix the socket outlet to its box.

If the box is a flush metal box sunk into the wall it is necessary to chop out some of the plaster at the knockout holes. If there is no room for an additional cable it will be necessary to knock out another blank and fit a PVC grommet into the hole before threading in the new cable. It is sometimes possible to take out the blank and fit a grommet without removing the box, but it is usually more satisfactory to remove the box first. Chop out a chase for the cable, lay the cable in the box and resecure the box to the wall. Make good the cable chase with plaster filler and allow the plaster to set before making the wiring connections, which are the same as for a surface-mounted socket outlet.

Socket outlets

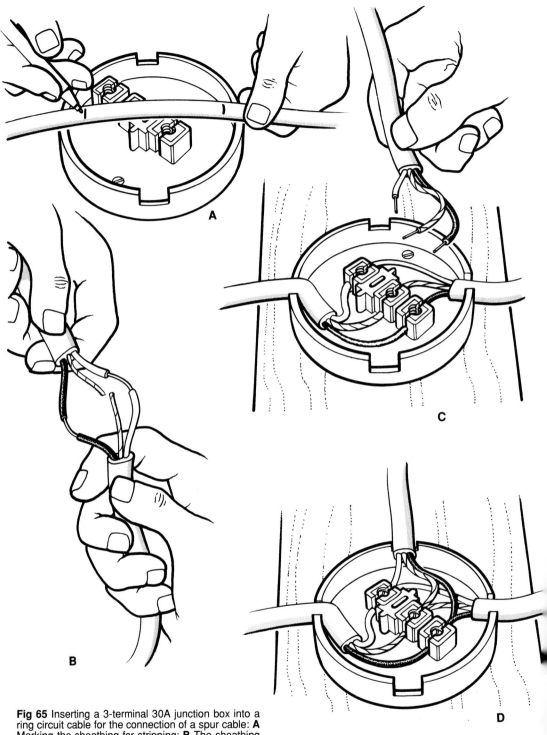

Fig 65 Inserting a 3-terminal 30A junction box into a ring circuit cable for the connection of a spur cable: **A** Marking the sheathing for stripping; **B** The sheathing stripped and some insulation removed from live and neutral wires; **C** The earth wire is cut and covered with green/yellow sleeving and all three wires laid in the terminals; **D** The three wires of the spur cable are connected to their appropriate terminals.

Fixing and wiring a junction box

Taking the supply from a junction box is a good method to choose when access to the underfloor space is easy and the cable to the new socket outlet can conveniently be run under the floor.

Lift floorboards and locate the ring cable. The most likely position is in close proximity to a socket outlet. At a convenient spot in the cable run where a junction box can be inserted into the cable, fix a piece of 100mm × 25mm timber between two joists.

Unscrew the cover of the junction box and fix the box to the timber. Place the cable over the junction box but do not cut the cable. With a ballpoint pen mark the cable at the centre of the junction box. Strip off the outer sheathing for about 50mm each side of the ink mark making 100mm of sheathing removed. Remove the terminal screws of the junction box. Strip about 5mm of insulation from the centre of the red and of the black cable. Place these wires over the outer terminals of the junction box with the bared section in contact with its terminal. Cut the earth wire in the centre; slip green/yellow PVC sleeving over each wire and place them over the centre terminal of the junction box.

Now take the end of the cable running from the new socket outlet. Strip off about 100mm of outer sheathing. Place this cable in the junction box with the respective wires over their terminals: red to red, black to black and earth wire to earth wire.

Trim the ends of the wires to fit, and strip off about 10mm of insulation from the red and from the black wire. Slip green/yellow PVC sleeving over the end of the earth wire. Place the wires in the terminals of the junction box on top of the existing wires. Replace the terminal screws and tighten them. Replace the junction box cover.

Extending the ring

Another way to add extra socket outlets to an existing ring circuit is to break into the ring at two adjacent socket outlets and to add a new loop of socket outlets, starting at the first old socket outlet and finishing at the second. Removing the cable linking the two existing socket outlets then creates a new enlarged ring, the only rule for which is that it must not serve an area larger than 100 square metres.

Extending the ring like this is often more convenient and less disruptive than adding spurs and may be the only answer if your socket outlets are already doubles, and have the maximum number of spurs, and you still need more socket outlets. Sadly, even modern builders rarely put in an adequate number of socket outlets – see the socket outlet provision recommendations in the table on page 177.

Extending a ring circuit is a similar job to adding spurs in that you need to add an extra cable to two existing socket outlets and then take the cable to new socket outlets as already described. You *must*, however, remove (or disconnect and cut back) the cable between the two existing socket outlets which means identifying two adjacent socket outlets on the ring and the cable which joins them.

The way to do this is to turn off the electricity and disconnect all the red and black cables at existing socket outlets. To test whether two socket outlets are adjacent, twist the red and black wires of one cable in one socket outlet together and then go to the second socket outlet to see whether you can get a continuity reading between a black and red wire in one of the cables. If not, try the other cable or another socket outlet. A word of warning – the *first* and *last* socket outlets on the ring will give this continuity reading and if you disconnect the cable which appears to join these, you will disconnect the ring from the mains supply!

Once you have established two adjacent (and convenient) socket outlets, run a new cable into the back of each socket outlet, connected to the live, neutral and earth terminals, and from these wires run the new socket outlets as an extra loop. Double check which of the two cables in each box is the one which joins the two socket outlets and remove it or cut it back.

8: Small fixed appliances

Many small electrical appliances in the home are fixed – that is, they are secured in place – and should be connected permanently to the house wiring. Typical appliances include wall heaters, extractor fans, cooker hoods, waste disposal units, water softeners, central heating and undercupboard lights. The way to connect these to the house wiring is via a *fused connection unit*.

Any of these appliances could, of course, be supplied from a 13A fused plug and socket outlet; however this method is not always satisfactory as the plug can be removed, a portable appliance used from the socket 'temporarily', with the result that the user forgets to replace the original plug. Also no socket outlet may be installed in a bathroom except that of an approved shaver supply unit containing an isolating transformer.

Any fixed appliance having an electrical loading of up to 3000 watts can be run from the ring circuit via a fused connection unit instead of from a socket outlet.

Fused connection units

A fused connection unit, originally termed a fused spur box, is basically a fuse unit containing a cartridge fuse as in a 13A fused plug. The fuse unit is mounted on a 'square' 1-gang faceplate which fits the standard l-gang mounting box, of either the moulded-plastic surface type or the metal flush-fitting version. Fuses used in fused connection units are the same size and rating as those used in plugs: primarily 3A and 13A, though 2A, 5A and 10A may sometimes be used.

The fused connection unit comes in a variety of types: unswitched, with and without flex outlet and with and without neon indicator; switched, also with and without flex outlet and with and without neon indicator. The standard model (see opposite) has a moulded plastic faceplate but versions are also available with metal faceplates in various finishes to suit the decor of the room concerned.

Function of a fused connection unit

The primary function of a fused connection unit is to connect a fixed appliance or fixed apparatus to a ring circuit or to a multi-outlet radial power circuit. Each appliance connected to either circuit must be protected by an individual fuse which with a plug and socket is the plug fuse; for a fixed appliance the fuse is in the fused connection unit. The current rating of the fuse is related to the current rating of the flex or cable connecting the appliance to the unit and also to the load (watts rating) of the appliance. Without the individual fuse the only protection would be the 30A circuit fuse. A switched fused connection unit also serves as an isolating switch where necessary, the switch being double-pole and therefore completely isolating the appliance from the mains.

Fused connection units may also be used for providing fused spurs from a ring circuit to feed a socket outlet or flex outlet unit.

Wiring a fused connection unit

Most fixed appliances have, or can have, a flexible cord for the connection to the unit. Others are connected by a cable, usually ordinary house-wiring PVC-sheathed cable, sometimes heat-resisting cable.

Where an appliance is fitted with flexible cord the fused connection unit must have a flex outlet, either a hole in the front or a knock-out at the bottom of the faceplate, if the flex is fitted directly.

Switched fused connection unit

A switched fused connection unit, with or without neon indicator has two pairs of terminals plus earth terminals. One pair of 'L' and 'N' terminals are marked 'FEED' (or 'SUPPLY' or 'IN') depending on the make (and age) of the unit. The other pair of 'L' and 'N' terminals are marked 'LOAD' (or 'OUT').

The wires of the circuit cable are connected to the FEED terminals, whilst the

Photo 61 *(above)* Unswitched fused connection units with and without front flex entry holes and neon indicators (*MK*).

Photo 62 *(below)* Switched fused connection units with and without front flex entry holes and neon indicators (*MK*).

124 Small fixed appliances

Photo 63 Rear view of switched fused connection unit showing terminals (*MK*).

flex (or cable) going to the appliance is connected to the LOAD terminals, the positions of the terminals varying with the particular make of the unit.

Unswitched fused connection unit

The arrangement of the terminals of an unswitched fused connection unit is similar to the switched type. Unswitched fused connection units are mainly of use in bathrooms, where no switch is allowed within reach of anyone using the bath or a shower (except a ceiling-mounted pullcord switch).

Circuits for fused connection units

A fused connection unit can be connected to the ring cable of a ring circuit in the same way as a socket outlet, the ring cable looping in and out of the FEED terminals of the connection unit. Alternatively the fused connection unit is connected on the end of a spur cable (see Fig 62), this being the method generally used for supplying the various fixed appliances such as sink waste disposal unit, heated towel rail and other such apparatus added later to an installation or where one cable, not two, is more convenient to the unit. For the spur, 2.5mm^2 2-core and earth cable is used.

Where a fused connection unit feeds a socket outlet or flex outlet unit, the cable between the unit and the outlet can be 1.5mm^2 2-core and earth.

Fixing a fused connection unit

A fused connection unit can, as stated, be mounted on either a moulded plastic surface box, preferably used with mini-trunking, or a metal box sunk flush into the wall.

For surface mounting remove a section of the thin plastic from the base of the box. Mark the position of the box on the wall, then drill and plug holes for fixing screws. Thread in the end of the sheathed cable and fix the box using No 8 wood screws.

Photo 64 Rear view of switched fused connection unit with flex outlet. Note the cord clamp (*bottom right*) (*MK*).

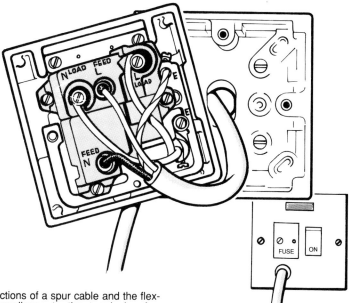

Fig 66 The connections of a spur cable and the flexible cord of an appliance to the terminals of a switched fused connection unit. This version has a neon indicator but the connections are identical to those having no neon indicator.

For flush mounting, cut a hole in the wall to take the box and a chase for the cable(s). Knock out a blank from the side of the box and fit a PVC grommet to protect the cable. Drill and plug fixing holes in the walls and secure the box with No 8 screws. Thread in the cable and make good the wall with filler before wiring up.

For flush fitting to partition walls, use a special type of box – see pages 85 and 86.

With either type of box the method of connecting the wires is the same.

Connecting the unit

Prepare the end of the spur cable by stripping off about 100mm of outer sheathing and about 10mm of insulation from the ends of the red and the black wires. Slip green/yellow PVC sleeving over the end of the bare earth conductor. Strip about 75mm of sheathing from the end of the appliance's flexible cord and about 10mm of insulation from the ends of each of the three conductors.

As we are dealing here with a switched version, the connection unit has four terminals plus earth terminal. Connect the brown wire to the LOAD 'L' terminal, the blue wire to the LOAD 'N' terminal and the green/yellow wire to the 'E' terminal. If the unit has a cord clamp, place the end of the flex sheath under the clamp and tighten the screws.

Connect the red wire of the circuit cable to the remaining 'L' terminal (marked 'SUPPLY' or 'FEED') and connect the black circuit wire to the corresponding and remaining 'N' terminal. Connect the green/yellow sleeved earth wire to the 'E' terminal of the unit.

Place the wires neatly in the box and fix the unit to the box using the two screws supplied.

Where a fused connection unit is supplied from the ring cable instead of from a spur cable, there are two pairs of red and black wires, these being connected to the respective 'L' and 'N' terminals plus an extra earth conductor. Otherwise the connections are the same.

Fixed electrical appliances

The fused connection unit supplying a fixed appliance is normally fitted close to the appliance and connected to it by means of a short length of flexible cord.

126 Small fixed appliances

Photo 65 Kitchen or bathroom wall fan heater fitted with programmable timer and thermostat (*Dimplex*).

Wall heaters

For fixed wall-mounted heaters, the fused connection unit is positioned a few inches above the skirting board, except in a bathroom where special arrangements are required as described later in this chapter. For a high-level radiant heater, including one in the bathroom, the connection unit is also fixed at a high level and at the side of the heater.

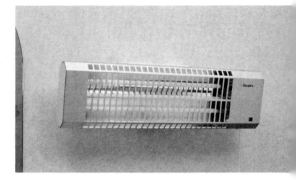

Photo 66 Infra-red wall heater with safe pull-cord switching (*Dimplex*).

Photo 67 Wall-mounted panel heater with programmable timer and built-in thermostat (*Creda*).

Base unit heaters

In the kitchen, there is often insufficient room to fit a radiator and a base unit heater, fitted into the plinth of a kitchen unit, is extremely useful. You can get both all-electric versions and versions which run from the central heating pipes, but fitted with an electric fan to blow warm air out into the room. The all-electric type shown here has its own fused control switch panel fitted above the worksurface with special cable run down the wall to an intermediate connector unit behind the heater into which the 6-core heater flex is wired.

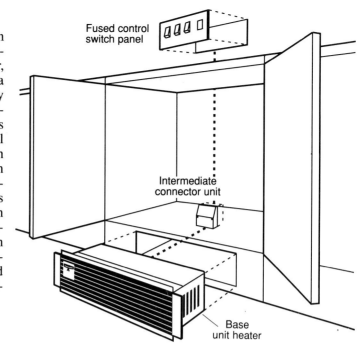

Photo 68 Wall-mounted control for base unit fan heater with three settings, built-in thermostat and thermal protection device (*Dimplex*).

Fig 67 Connections to an electric base unit heater. The fused control switch panel is wired to the ring circuit (some designs require a switched fused connection unit first) with cable run to the intermediate connector unit, from where flex is run to the heater.

Waste disposal unit

You need control of a sink waste disposal unit above the worktop, so fit the switched fused connection unit here with 1.5mm^2 cable going to a flex outlet unit close to the waste disposal unit (see page 136).

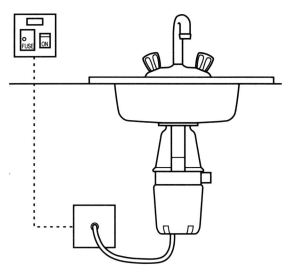

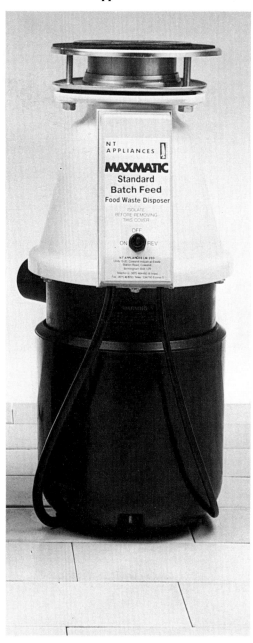

Fig 68 Sink waste disposal unit connected to a flex outlet unit under the sink and controlled by switched fused connection unit above the sink.

Extractor fan and cooker hood

For an extractor fan, fit the fused connection unit as close as possible to the fan, with any special control switch at a lower level. With extractor fans in bathrooms and toilets with inadequate ventilation, the wiring for the fan must be through the light switch.

The fused connection unit for a cooker hood can either be to one side of the hood or at a lower level, just above the worktop.

Central heating

The fused connection unit supplies the pump, boiler and the other electrics of a central heating system. Fix it close to the boiler, programmer or junction box as convenient.

Point-of-use water heater

For a kitchen sink or bedroom basin water heater, the fused connection unit is fixed near

Photo 69 Waste disposal unit (*NT Appliances*).

Small fixed appliances 129

Photo 70 Typical window-mounted extractor fan (can also be wall-mounted) (*Xpelair*).

Photo 71 Recessed 'Loovent' fan for bathrooms or toilets having inadequate ventilation (*Airflow*).

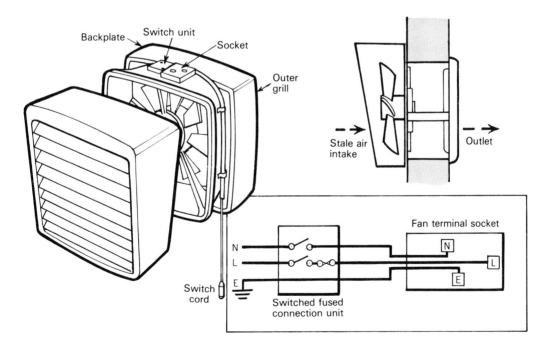

Fig 69 *Above*: A typical extractor fan. *Top right*: The fan fitted into a wall, but most fans can also be fitted into a hole cut out of a window pane. *Right*: The fan is supplied from a switched fused connection unit – note that some fans are double-insulated and so do not require the earth connection.

130 Small fixed appliances

Photo 72 Cooker hood fitted between adjacent wall cupboards (*Xpelair*).

the flex outlet of the heater. For a bathroom water heater, an unswitched fused connection unit is fixed outside the bathroom and the water heater is controlled by a cord-operated switch, though where the water heater is of the instantaneous type the switch is used only as an isolating switch.

Shower pumps

An electric shower needs its own separate circuit run all the way from the consumer unit – see Chapter 10. But a shower pump, fitted to boost an existing shower or to provide a 'power' shower, can be supplied from a switched fused connection unit wired into the local ring circuit.

The fused connection unit must be fitted outside the bathroom, from where cable is taken to a flex outlet unit (*see Bathroom towel rail*) to connect to the pump flex.

Water softener

A water softener, typically fitted under the kitchen sink, needs an electric supply to power the clock which operates the automatic regeneration system. It can be wired in the same way as a waste disposer, with a fused connection unit above the worktop and a flex outlet unit below.

Undercupboard lighting

In a kitchen, undercupboard lighting will often be run from a switched fused connection unit wired into the kitchen ring circuit, although the lights themselves will usually have their own switches.

Bathroom towel rail

An electrically heated towel rail is of necessity fixed to the bathroom wall at a low level, as is a bathroom heater whether it incorporates a towel rail or not. This means that the flexible cord of the towel rail or other low-mounted heater cannot be connected to a fused connection unit. Instead the flex is connected to a flex outlet unit fixed near the appliance.

A flex outlet unit is a 'square' moulded

Small fixed appliances 131

Photo 73 Wall-mounted boiler for tea making or providing hot water (*Creda*).

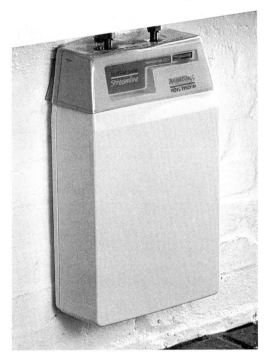

Photo 74 Point-of-use water heater can be mounted either under or over sink (*Heatrae Sadia*).

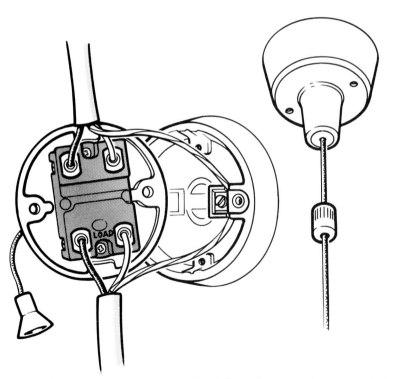

Fig 70 The cable connections at a 16A double-pole cord-operated ceiling switch. The earth conductors are connected to an earth terminal in the pattress box.

132 Small fixed appliances

plastic faceplate having a flex outlet hole in the centre and three terminals situated at the back of the plate for the connection of the live, neutral and earth conductors respectively of the flexible cord of the heater and the (1.5mm^2) cable running from the switched fused connection unit outside the bathroom or from a ceiling-mounted cord-operated double-pole switch situated inside the bathroom (wired to an *unswitched* fused connection unit).

A flex outlet unit, also used with some wash-basin and sink water heaters and other appliances where the isolating switch is of the cord-operated type and remote from the appliance, is mounted on a 1-gang box, which can be either the surface-mounted type or the flush-fitting version.

Shaver supply unit

A shaver supply unit has a 2-pin socket outlet which will accept razor plugs to British, American, Continental and Australian standards. The unit incorporates a double-wound isolating transformer to provide an earth-free mains voltage supply to the socket. It has a self-resetting thermal overload device to prevent any appliance (or lamp) being used from the socket other than a shaver. Most models are dual voltage for 240V and 115V.

The shaver supply unit is the only socket outlet permitted in a bathroom. As the unit is made to the exacting requirements of British Standard BS3052 it may be situated in any position in a bathroom and can be supplied direct from a ring circuit spur cable without the intervention of a fused unit. Alternatively the unit may be supplied from a lighting circuit.

Shaving strip lights are also available fitted with a shaver supply unit, and provided the unit is made to BS3052 it may be installed in a bathroom.

Shaver socket outlet

A shaver socket outlet is identical to the shaver supply unit except that it has no isolating transformer. It must *not* therefore be installed in a bathroom, but it may be installed in other rooms. The circuit for the

Photo 75 *Left*: Shaver socket outlet suitable for use in bedrooms and in rooms other than bathrooms; *Right*: Shaver supply unit, with isolating transformer, suitable for use in bathrooms (*MK*).

Small fixed appliances 133

Photo 76 Oil-filled electric towel rail, wired to flex outlet unit (*Dimplex*).

Photo 77a Front view of a flex outlet unit which fits a 1-gang mounting box; **Photo 77b** Rear view of unit showing terminals and cord clamp (*MK*).

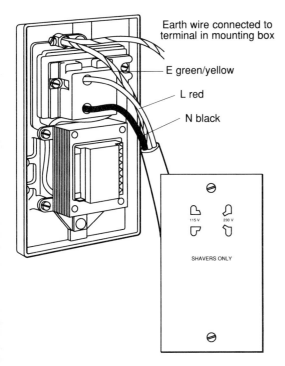

Fig 71 Rear view of a bathroom shaver supply unit with the spur cable connected to the terminals. The mounting box may be either a surface-mounted or a flush-fitting version.

shaver socket outlet can be a fused spur cable or the socket may be supplied from a lighting circuit.

Some shaving mirror lights are fitted with this type of socket and must not be installed in a bathroom.

Double-pole switches

A double-pole switch operates on both the neutral and live sides of the supply thus completely isolating whatever is 'downstream' from the mains. A light switch, on

134 Small fixed appliances

the other hand, interrupts only the live side of the mains supply.

Switched fused connection units incorporate a double-pole switch, but where an unswitched unit is used (as for example in a bathroom) a double-pole switch can be used later in the circuit – in this instance a ceiling-mounted pull-cord switch.

Wall-mounted double-pole switches are also available and can be used for connecting spurs to sockets for appliances like washing machines – see page 136.

Clock connectors

A clock connector performs the same function as an unswitched fused connection unit, but is fitted with a 2A fuse.

Originally, clock connectors were designed for connecting electric clocks to the mains. If you have an old clock, you can still use a clock connector like this but these days most electric clocks are battery operated.

Clock connectors do, however, have an additional use, which is to connect low-wattage appliances to the lighting circuit rather than to the ring main circuit. The clock connector is wired into the lighting circuit (connected as a spur from a convenient junction box or loop-in ceiling rose) and the flexible cord of the appliance is connected to it.

The maximum wattage that a 2A clock connector can carry is 480W and typical examples of their use are extractor fans and shaver lights or supply units in the bathroom where there are no socket outlets and running connections from the local lighting circuit will be a lot easier.

The circuit cable is connected into the back of the clock connector whilst the appliance flex is connected into the fused 'plug' which, when removed, will isolate the appliance.

Photo 78 Shaver striplight suitable for use in bathrooms (*MK*).

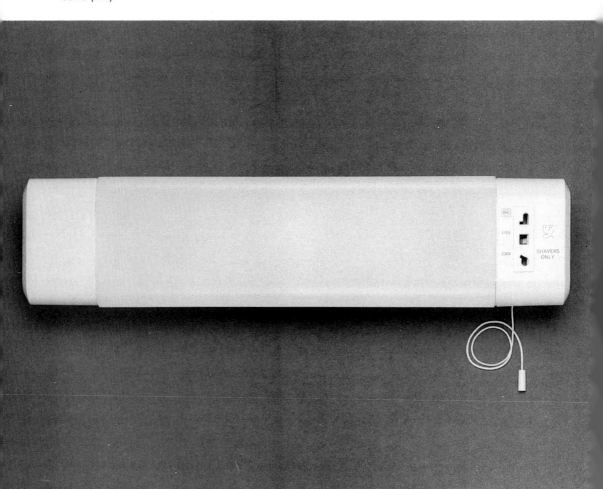

Small fixed appliances

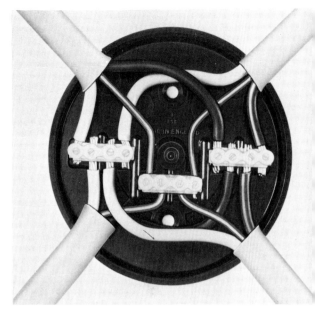

Photo 79 Fused clock connector showing cord outlet for the flex of a mains-powered electric clock (*MK*).

Photo 81 Spur cables supplying fused connection units can be connected to the ring circuit cable using this type of 30A junction box (*Ashley & Rock*).

Photo 80 Detachable section removed showing 2A fuse (*MK*).

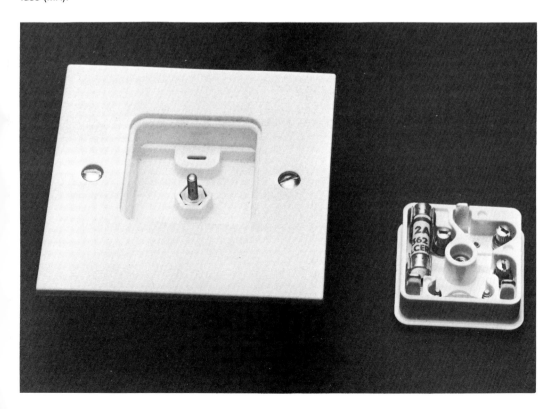

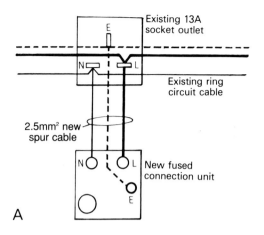
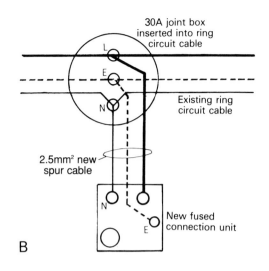

Fig 72 Connecting a spur cable supplying a fused connection unit to the ring circuit cable: **A** To the terminals of a 13A socket outlet; **B** To a 30A junction box inserted into the ring cable.

Lighting from fused spurs

It is often more convenient from the wiring aspect to supply new lights from a fused spur branching off the ring circuit than to break into a lighting circuit. This applies especially to wall lights, and outside lights.

If an unswitched fused connection unit is used, cable is taken from this to a junction box and thus to a light switch and light fitting in the normal way. If a *switched* fused connection unit is used, cable can be taken directly from this to the light fitting with the fused connection unit also acting as the light switch. In both cases, a 3A fuse should be fitted into the connection unit.

Connecting spur cable to ring circuit

The spur cable supplying the fused connection unit can be connected to the ring circuit either at the terminals of a ring socket outlet or at a 30A junction box inserted into the ring cable – see Fig 72.

The size of cable used for the spur must be 2.5mm² 2-core and earth since it will be protected only by the circuit fuse so must be capable of carrying 30A. The size of cable used *after* the fused connection unit (to an appliance, socket outlet or flex outlet unit) can be 1.5mm², since it will be protected by the (maximum) 13A fuse in the fused connection unit.

Fixed appliances under worktops

The socket outlets and fused connection units in a modern kitchen with continuous worktops will typically be just above worktop level.

This creates a problem for connecting appliances like washing machines, tumble driers and dishwashers since a) you want to be able to switch them on and off (perhaps in an emergency) from above the worktop and b) you want to be able to remove the appliance for cleaning the floor.

The answer here is a wall-mounted double-pole switch (or a fused connection unit) with a neon indicator, fitted above the worktop, wired into the kitchen ring circuit, with a cable buried in the wall leading down to a single unswitched socket outlet in the wall below the worktop and behind the appliance. An alternative would be a fused connection unit with cable leading to a flex outlet unit (as shown in Fig 68), but this would not make it so easy to remove the appliance. If a double-pole switch is used for the spur, the cable must be 2.5mm² 2-core and earth; with a fused connection unit, 1.5mm² cable can be used.

Washing machines (or dishwashers or tumble driers) *not* under worktops can be connected in the normal way via plugs to socket outlets, most conveniently just above skirting board level.

9: Bells, buzzers and chimes

Bells

The ordinary domestic bell is normally powered either by batteries or by a mains-supplied safety isolating transformer. The combined voltage of the batteries is typically 6V; transformers come either with fixed outputs (typically 8V at 1A or 0.5A or 12V at 1A) or with a choice of outputs (typically 3, 5 and 8V at 1A). The output you need depends on the design of the bell and, in some instances, the length of the circuit wires.

The most commonly used bell is the trembler type, so named from the trembling action of the striker mechanism. A trembler bell can be used on a DC (battery) supply or from an AC mains-supplied safety isolating transformer.

Buzzers

A buzzer which produces a buzzing note instead of the tinkle of a bell operates similarly to a bell and has the same basic mechanism, the principal differences being that the buzzer has no striker and no gong.

Bell (or buzzer) circuit

This is a simple circuit which includes a power source (battery or transformer), a bell (or buzzer) and a bell push. These are connected together by wire, termed bell wire which is of lightly insulated low voltage rating, usually twin-core. The wire is fixed to walls and to skirting boards and picture rails with cable clips or white-painted nails driven between the two cores of twin bell wire. An alternative is to tack the bell wire to the surface with blobs of adhesive from a hot-melt glue gun.

Bell pushes

Bell pushes are made in a variety of styles and types, including illuminated models.

A bell push basically consists of two contacts, one of which is spring loaded to

Photo 82 Plastic bell push with hidden fixing screws (*Friedland*).

keep the contacts separated, and a cover which contains the push button. The two circuit wires are connected to terminals on the contacts. When the button is depressed the contacts close and complete the circuit, causing a flow of current which operates the bell (or buzzer or chimes).

An illuminated push contains a small bulb connected to the two contacts. This completes the circuit through the coils of the bell (or buzzer or chimes) and lights the bulb at all times. When the push is pressed the light goes out as the contacts are shorted, but as soon as the button is released the bulb lights again.

As an illuminated push consumes energy continuously it can only be fitted into a circuit powered from a transformer. If battery powered the battery would be exhausted in a couple of days.

138   Bells, buzzers and chimes

Photo 83 Popular patterns of door chimes (*Friedland*).

Door chimes

Door chimes usually consist of two bars or tubes and a spring-loaded plunger in a solenoid. When the bell push is operated the solenoid is magnetised and draws the plunger into the solenoid with a force which causes it to strike one of the tuned bars or tubes. When the push is released and the solenoid de-energised the spring returns the plunger but with a force which causes it to strike the second tuned bar or tube so producing a 'ding-dong' double note.

Door chimes, as with most bells and buzzers, are normally operated from a push at the front door. Some chimes, with the choice of one or two notes, can also be operated from the back door. The front door push gives the double note but operating the push at the back door gives a single note only, so informing the householder of the whereabouts of the caller.

As well as 'ding-dong' chimes, you can also get warbling chimes with repeated signals (especially good for the hard of hearing) and electronic chimes which play a choice of well-known tunes. The latest developments include portable door chimes and door phones, which allow the caller to be identified before opening the door.

Power sources

Nearly all bells and most chimes have a purpose-made compartment inside to take four or more 1.5V batteries and with two terminals for connection to the circuit wiring. Long-life batteries should be used wherever possible.

Buzzers and some bells and chimes need an external power source – either batteries or a mains-operated safety isolating transformer. Some chimes will operate only from a transformer (which is supplied); others only from a battery.

Installing a bell or chimes transformer

Transformers for bell, buzzer and chime circuits are designed to supply Safety Extra Low Voltage (SELV). The transformers are normally double insulated which means they require no earth connection.

The majority of chimes will need a transformer with an 8V 1A output, whilst the majority of bells need an 8V 0.5A output. For circuits supplying multiple chimes (or extra long circuits where voltage drop might be a problem), a transformer with a 12V 1A circuit is needed. In all cases, always ensure that your transformer is compatible with the bell, buzzer or chimes: the manufacturer's instructions will supply the information required.

The transformer can be connected directly to a lighting circuit, using 1.0mm^2 two-core and earth PVC-sheathed cable; with a double-insulated transformer, the earth wire should be cut back.

Alternatively the transformer can be supplied from the ring circuit via a fused connection unit fitted with a 3A fuse. A 2.5mm^2 2-core and earth PVC-sheathed cable is run from the point of connection with the ring circuit (see Figure 62). This spur cable is run into an unswitched fused connection unit conveniently sited, and from the LOAD terminal of the unit a 1.0mm² 2-core and earth PVC-sheathed cable is run to the mains terminals of the transformer, cutting the earth wire of the cable back. The low-voltage bell wire is connected to the output terminals of the transformer.

Extensions

When the main electric bell or buzzer cannot be heard in remote parts of the house it is possible to run an extension bell from the main bell (or buzzer). So that both bells operate simultaneously and effectively it is necessary that (i) the circuit is supplied from a mains transformer, (ii) the bells are identical, (iii) the bells are suitable for use with a transformer.

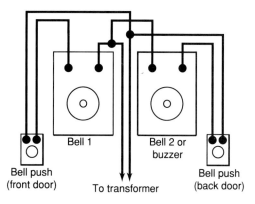

Fig 73 Two bells with separate pushes for the front and back door respectively, wired on separate circuits from a single power source.

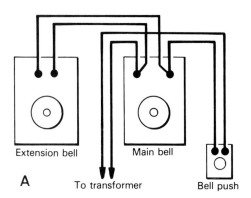

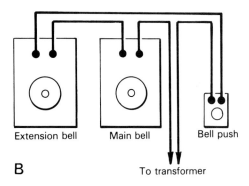

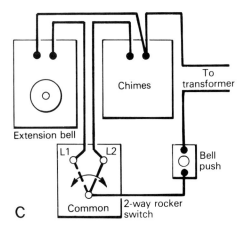

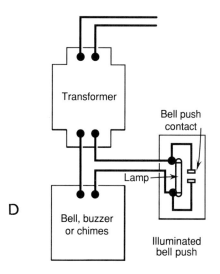

Fig 74 Bells and chimes circuits: **A** An extension bell connected in parallel with the main bell; **B** Extension bell connected in series with the main bell; **C** Extension bell and chimes controlled by a change-over switch so that either can be switched into circuit; **D** Chimes, buzzer or bell powered from a transformer which also allows the push to be of the illuminated type.

140 Bells, buzzers and chimes

There are two methods of connecting the two bells: (i) connect them in parallel, or (ii) connect them in series.

For parallel connection, a twin wire is run from the extension bell and is connected to the terminals of the existing bell along with the existing wires which are not disturbed.

For series connection, run the same twin wire from the extension bell to the existing bell. Disconnect one of the existing wires and in its place connect one of the new wires. Connect the old wire to the other new wire in the form of a joint and cover it with PVC insulation tape. There is a third method which is more effective but allows only one bell to operate at a time. A change-over switch is inserted into the wiring, this being a normal 2-way lighting switch. The faceplate should be marked MAIN BELL and EXTENSION BELL. Whenever it is not being used in the EXTENSION position it is important that the bell is switched back to MAIN BELL.

It is possible to run an extension bell or buzzer from door chimes by connecting the two wires from the extension bell to the terminals of the chimes, but the extension bell rings only when the caller's finger remains on the push, so the short response from the bell when the chimes produce the single 'Ding' note may not be heard.

It is even more difficult and usually impossible to get successful operation of two chimes connected in parallel on the one circuit. Two identical chimes wired in series should work, but you will need a transformer with an output of 12V at 1A.

Portable door chimes

A portable door chime consists of two units: a transmitter, which is connected to the bell push in the normal way, and a receiver which can be mounted on the wall or carried around the house. Communication between the transmitter and the receiver is by radio waves and both transmitter and receiver are fitted with batteries.

A portable door chime can be used as the only chime in the house or can be fitted as an extension to an existing transformer-powered chime.

An alternative use for a portable door chime is as a low-cost calling device. The transmitter has a push button on it which will sound the chime when pressed.

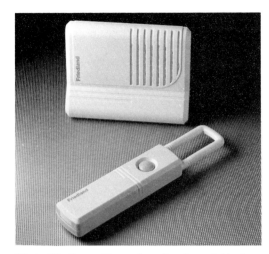

Photo 83a Transmitter and receiver for portable door chimes (*Friedland*).

Door entry telephone systems

There is a good choice of 'intercom' systems which allow the home owner to talk to whoever is at the door before opening it to let them in.

Some systems, primarily designed for flats, have a simple push button at the front door which rings a telephone in the flat. When the identity of the caller is established, a button on the internal telephone handset releases the door lock.

A simpler version can be used to replace an existing bell or chime. When the bell push

Photo 83b Door entry intercom replaces existing door chimes and push (*Friedland*).

Bells, buzzers and chimes 141

on the external microphone unit is pressed the internal chimes sound and, by pressing on the front of the chimes unit, a conversation can be held with the caller.

More sophisticated intercom systems use tiny video cameras so that you can see who is at the door as well as talk to them.

Burglar alarms

There are numerous burglar-alarm systems available for installation in the home and for the do-it-yourselfer there are kits available. It is best to use a basic kit which can be extended as required.

A typical do-it-yourself burglar alarm kit consists of: (i) a key-operated or touch-pad control box which is the nerve centre of the system and which is fitted with an anti-tamper switch; (ii) a rechargeable battery; (iii) magnetic contacts, which can be fitted to doors and opening windows; (iv) pressure pads for placing under the carpet on the stairs or in front of valuable objects; (v) an alarm bell or siren for fitting outside the house; (vi) passive infra-red sensors which detect movement within the house; (vii) six-core cable for wiring the system.

Extra items include a strobe light for the outside bell unit, 'personal attack' buttons, dummy bell boxes and additional magnetic sensors.

Power supply

Although a burglar alarm can be operated from a battery, it is much better to power it from the mains with an internal rechargeable battery in the control box (and sometimes in the external bell unit as well) as a back-up in the event of power failure. This means that the alarm will always work. Using 6-core alarm cable will mean that any attempt to cut the wire will result in the alarm sounding – only two cores are used to link most protective devices, whilst the others form a continuous circuit which will set the alarm off if broken.

Installation

Full instructions should be provided with a do-it-yourself burglar alarm kit which will give advice on where best to site the sensors and detectors as well as giving detailed wiring instructions.

The low-voltage wiring is straightforward, consisting of making two or more connections at each detector or sensor and wiring the cable to the correct terminals on the control box and running a cable from the control box to the external bell unit. Some systems allow you to divide the house up into zones and there will be an adjustment for the time which the system allows you to leave through the front door after setting the alarm and the time allowed to turn it off after coming back into the house.

The connection to the mains should be from a fused connection unit, wired as a spur from a convenient socket outlet circuit – see page 136.

Fig 75 Typical low-voltage wiring diagram for simple do-it-yourself alarm.

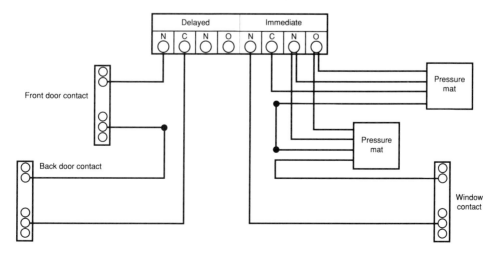

10: Cookers and showers

The two most powerful pieces of electric equipment in the home are a cooker (free-standing or built-in) and an instantaneous electric shower. Both need their own separate circuit.

Installing an electric cooker

An electric cooker, other than a table cooker having an electrical loading not exceeding 3000 watts, needs an exclusive circuit supplied from a fuseway in the consumer unit. No extensions may be made to a cooker circuit to supply any other outlet or apparatus, the only exception being a socket outlet embodied in the cooker control unit.

Circuit requirements

The circuit starts at a fuseway, usually in the consumer unit, and the cable runs to a cooker control unit fixed within reach of the person using the cooker. From the control unit a cable runs to the cooker. With a freestanding cooker it is usual to insert a cable outlet unit in this cable, fixed to the wall about 600mm above floor level. From the outlet unit a final length of trailing cable runs into the cooker terminal box. The trailing cable is needed to allow the cooker to be pulled away from the wall to facilitate cleaning. With a built-in oven or hob the cable from the control unit runs direct into the terminal box of the appliance.

Current rating of circuit

A free-standing cooker (or separate split-level oven and hob) can have a total loading of 12kW or more. This wattage implies a current of 50A, but in fact the cooker can be installed on a circuit rated at just 30A.

The reason for this is that not all of the cooker (oven and hob) will be used at *full* power at once (even when the Christmas lunch is being cooked) and the normal method for calculating the current rating for a circuit is to take the first 10A at 100 per cent and the remainder at 30 per cent, allowing 5A for a socket outlet included in a cooker control unit. For a 12kW cooker, this would give 10 + 12 + 5 = 27 amps, which is within the current carrying capacity of a 30A circuit.

It is, however, good practice to install a cooker circuit rated at 40A or 45A, so that it can cope with the whole cooker being used together with a 3kW kettle (plugged into the socket outlet in a cooker control unit) and so that it will always have the capacity for future occupants even if your cooker does not need this rating of circuit. Cooker control units and wall-mounted double-pole switches suitable for use with cookers are all rated at 45A, and 45A fuses and 40A or 45A miniature circuit breakers are now readily available for consumer units.

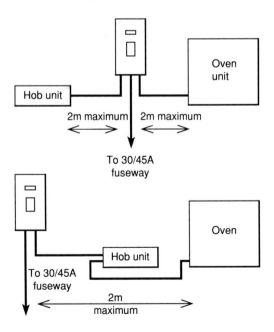

Fig 76 Alternative methods of connecting the two sections of a split-level cooker to the cooker control switch.

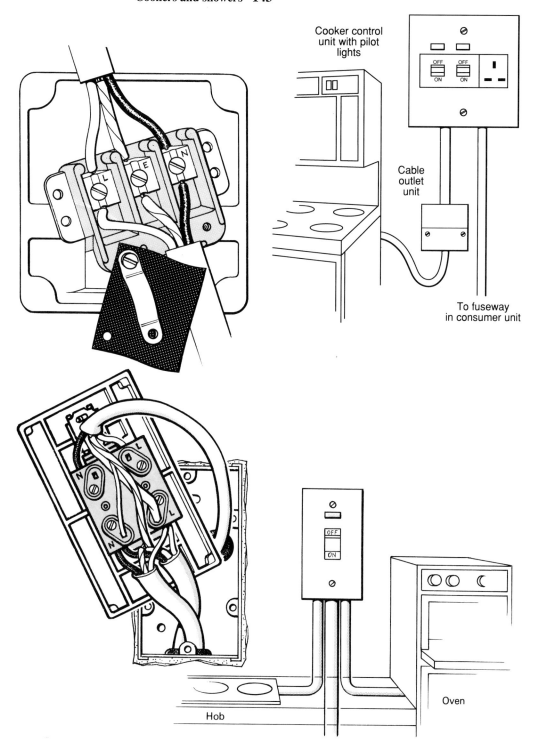

Fig 77 *Above*: Control unit and cable outlet arrangement with a free-standing electric cooker. *Left*: Connections at the terminal outlet box. *Below*: Wiring layout of a split-level cooker and showing the cable connections at the cooker control switch.

144 Cookers and showers

Circuit cable size

The size of cable you need for a cooker depends on the rating of the circuit (30A, 40A or 45A), the length of cable run and the method of fusing at the consumer unit. With cartridge fuses or miniature circuit breakers (MCBS), you can use 4mm^2 2-core and earth PVC-sheathed cable for 30A circuits and 6mm^2 cable for 40A or 45A circuits. With rewirable fuses or cable runs longer than around 20m, you should go up a size – ie 6mm^2 for a 30A circuit and 10mm^2 for a circuit rated at 40A or 45A.

Larger cable sizes are more difficult to handle and it would be worth considering changing to MCBS – especially if a new switchfuse unit is being fitted.

Materials required

The materials required for wiring a 40A or 45A cooker circuit are: a length of 6.0mm^2 2-core and earth flat PVC-sheathed cable; a length of green/yellow PVC sleeving; plastic cable clips for this size and type of cable; one cooker control unit (or cooker control switch); one cable outlet unit (for a free-standing cooker); a 45A cartridge fuse and unit or 40A or 45A MCB. For a free-standing cooker, the cable connecting the cooker to its control unit must be of the same size as the circuit cable; for a split-level cooker, the cable running to each section must be of the same size as the circuit cable although the total load of the cooker is divided between the two sections. Ideally the 45A fuseway will be a spare fuseway in the existing consumer unit. Where, as is usual, there is no spare fuseway, short of replacing the consumer unit by one having more fuseways it is necessary to install a switchfuse unit for the cooker circuit (see Fig 97).

Planning and wiring the circuit

First decide where the cooker control unit is to be situated. Normally this is at the side of the cooker, about 1.5m above floor level. The cooker control unit (or switch) should be within 2m of the cooker, but not positioned above the hob if a cooker control unit with a socket. In general, it is better to use a cooker control switch (a 45A double-pole switch) rather than a cooker control unit with a socket outlet, because of the danger of trailing flexes near the hot hob. Sufficient sockets should be provided in the rest of the kitchen. One cooker control unit may supply both sections – the hob and the oven – of a split-level cooker provided both sections are within 2m of the one control unit. Otherwise a separate control unit is required for each section, though both control units are connected to the one cooker circuit. Rarely are two required, for if the control unit is fixed midway between the sections, they can be up to 4m apart, not necessarily on the same wall – see Fig 76 for the two different methods of wiring.

A cooker control unit must be readily accessible at all times and must not, for instance, be situated under a work top or in a cupboard.

The route for the circuit cable depends largely on the structure of the ground floor. With a suspended floor, run the circuit cable down the wall from the consumer unit into the void beneath the floorboards and up the kitchen wall to the control unit. With a solid floor it is necessary to run the cable up the wall into the void above the ceiling (or in the roof space of a bungalow) and down the kitchen wall to the control unit. This cable can be run in mini-trunking on the surface or buried in the wall plaster.

The cable running down the wall from the control unit to the cable outlet unit for a free-standing cooker can again be fixed to the wall surface or buried in the plaster, whichever is desired. The cables running to the sections of a built-in split-level cooker can likewise be buried in the wall if preferred, but with built-in units it may be possible to hide the cables in the structure. However, in no circumstances should the cables be sunk into the plaster until all units have been fixed and there is to be no further drilling or plugging of walls.

Fixing the cooker control unit switch

Cooker control units and cooker control switches are available with and without neon indicators (two for a cooker control unit) in both surface and flush mounting versions. Some surface-mounted cooker control units come with their own box.

The first task is to remove the cable knock-outs from a plastic surface-mounting box or a metal flush-mounting box (fit PVC grommets to the holes) and to secure the box to the wall – on the surface or in a hole made in the wall respectively. For surface mounting, fit mini-trunking to the wall; for flush mounting, cut a chase in the wall plaster to take the cable.

Lay the circuit in the mini-trunking (or bury it in the wall and make good the plaster) and similarly bring the cable from the cable outlet unit (or from the built-in appliances) into the mounting box. Prepare the ends of the cables allowing around 200mm of the unsheathed portion for the connection to the unit or switch.

Connect the wires of the circuit cable to the MAINS terminals – red to 'L', black to 'N' respectively. Connect the wires of the cable running to the cooker to the corresponding LOAD terminals and connect the sleeved earth conductors to the 'E' terminal of the unit. With a split-level cooker there may be two cables for connection to the LOAD terminals, depending on the method of wiring adopted. Place the cables into the box, fix the unit and replace the outer cover.

Fixing a cable outlet unit

This connector unit consists of a metal frame containing a 3-way terminal block; a cable clamp and two fixing screws; a flush metal box and a moulded plastic cover plate (see Fig 77). Sink the box into the wall by cutting a hole. Remove a knockout blank and fix a grommet into the knockout hole. Thread in the cable running down from the control unit, prepare the end and connect the three wires respectively to the three terminals.

Take the cable running from the cooker and prepare the end. Connect the wires to the same terminals as the other cable: red to red, black to black and sleeved earth wire to sleeved earth wire. Secure the end of the sheathing of this cable under the cable clamp. Fix the metal frame to the box using the captive screws. Fix the cover plate using the screws supplied.

An alternative version of this outlet box has no terminal block. The cable from the control unit to the cooker passes through the box, is secured under the clamp, but it is not cut.

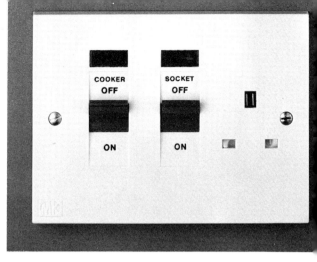

Photo 84 Flush-pattern cooker control unit with neon indicators (*MK*).

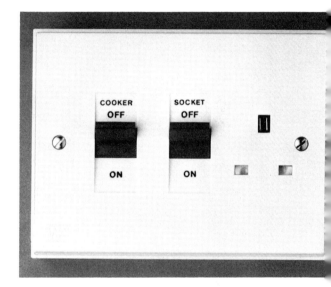

Photo 85 Surface-mounted cooker control unit with supplied mounting box (*MK*).

Photo 86 45A double-pole cooker control switch to fit in 1-gang box (*Ashley & Rock*).

146 Cookers and showers

Photo 87 45A double-pole switch suitable for cooker (self-adhesive labels supplied) for flush or surface mounting (*MK*).

Photo 88 Rear view of cooker control switch showing terminal arrangement (*MK*).

Photo 89 Cable outlet unit with plastic faceplate. *Centre*: for joining two cables; *Right*: for cable passing right through (*Crabtree*).

Cookers and showers 147

Photo 90 Modern slot-in cooker with hob, grill, oven and built-in microwave (*Creda*).

Installing an electric shower

An electric shower unit is basically an instantaneous electric water heater which heats the water as it flows over the elements. The unit operates on the simple principle that when the cold water inlet valve is opened the pressure of the water closes an electric switch, allowing the current to flow and heat the element which in turn heats the water before it flows out through an open outlet pipe. The temperature of the outflow of water varies with the rate of flow: the slower the flow the higher the temperature and vice versa.

As the water is heated only when it flows the electrical loading of the unit needs to be 7kW – some models have an 8.4kW loading – to obtain an adequate flow at the required temperature for showering.

The electric current demand for a 7kW shower unit is 29A, that of an 8.4kW shower unit 35A. This means that the current rating of a circuit supplying a shower unit must be at least 30A, and preferably 40A or 45A so that it can cope with more powerful showers.

Photo 91 Built-in ceramic hob remains cold to the touch even when on its maximum setting (*Creda*).

148 Cookers and showers

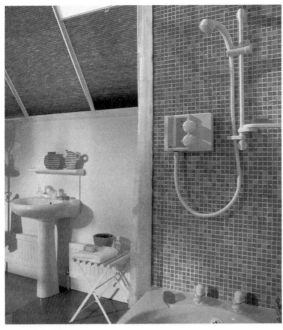

Photo 92 Two electric showers. *Above*: 8kW electric shower positioned over bath; *Right*: 8.4kW electric shower, with built-in temperature sensor and visual display (*Mira*).

Planning the circuit

A shower unit requires a double-pole switch to isolate it from the mains when necessary. The switch should be situated within reach of the shower unit, which means that a wall-mounted switch is unsuitable, therefore a 45A double-pole cord-operated ceiling switch must be fitted.

The circuit cable for a shower unit runs from the circuit fuseway in the consumer unit, or from a new switchfuse unit, to the cord-operated ceiling switch. From the switch a length of the same cable is run to the terminals of the shower unit.

Size of circuit cable

The size of cable you need for a shower depends on the rating of the circuit (30A, 40A or 45A), the length of cable run and the method of fusing at the consumer unit. With cartridge fuses or miniature circuit breakers (MCBs), you can use 4mm^2 2-core and earth PVC-sheathed cable for 30A circuits and 6mm^2 cable for 40A or 45A circuits. With rewirable fuses or cable runs longer than around 20m, you should go up a size – ie 6mm^2 for a 30A circuit and 10mm^2 for a circuit rated at 40A or 45A.

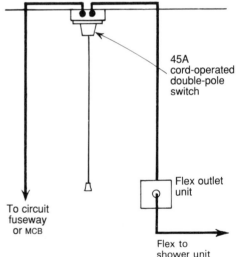

Fig 78 Circuit wiring for an electric shower unit.

Larger cable sizes are more difficult to handle and it would be worth considering changing to MCBs – especially if a new switchfuse unit is being fitted.

Materials required

The materials required for installing a 40A or 45A shower circuit are: 2-core and earth flat PVC-sheathed cable (6.0mm^2 or 10.0mm^2);

green/yellow PVC sleeving; one 45A double-pole cord-operated ceiling switch with neon indicator; one l-gang moulded plastic mounting box (if required); cable clips; wood screws and wall plugs. You may also need a switchfuse unit (see page 184) if there is no spare fuseway or MCB in the consumer unit.

Wiring the circuit

Plan the route of the cable from the unit to the switch – normally up the wall above the consumer unit into the ceiling void, between and across joists which are drilled to accommodate the cable and into the bathroom. As the cord-operated switch is fixed to the ceiling of the bathroom the circuit cable should be run up into the roof space to the ceiling switch. To conceal the cable run it up the wall inside the airing cupboard or other cupboard, provided this would not substantially lengthen the cable run and thus increase the cost. Alternatively the cable can be buried in the plaster or enclosed in mini-trunking.

From the ceiling switch run another length of the same cable down the bathroom wall and into the shower unit. This cable also can either be buried in the wall or enclosed in mini-trunking rather than fixing it to the wall surface using cable clips. If the bathroom is already tiled use mini-trunking, but if yet to be tiled it is as well to bury the cable.

Before installing the circuit cable, pierce a

Photo 93 8kW electric shower with lower power setting for summer use (*Heatrae Sadia*).

small hole in the ceiling at the point where the ceiling switch is to be fixed. Go into the roof space and check whether the tiny hole is next to a joist or whether it could be moved a few inches to bring it against a joist for fixing the switch. If a joist cannot be used for fixing, insert a piece of wood batten between the joists with a hole drilled in it for the two cables. Pierce a hole in the ceiling at the point where the other cable is to run down the wall

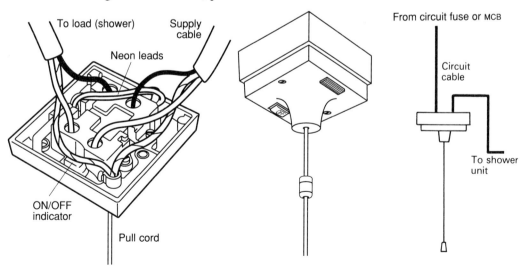

Fig 79 Cable connections at a 45A double-pole cord-operated ceiling switch with neon indicator and mechanical flag indicator. The switch can be used as the isolating switch for a shower unit.

150 Cookers and showers

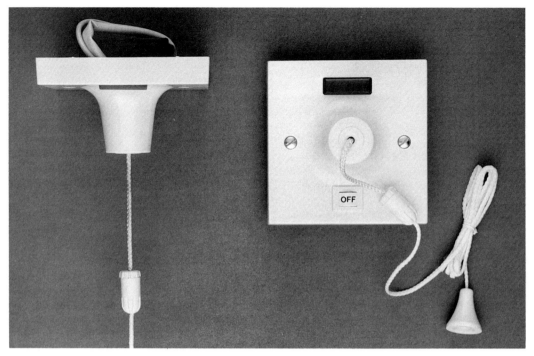

Photo 94 45A double-pole ceiling switch with neon and with mechanical flag indicator (*MK*).

to the shower unit. Install the circuit cable along its prepared route and pass the end through the hole in the ceiling. Cut the cable to length allowing about 250mm for connections. Using a ballpoint pen mark the sheathing at the switch end of this cable: MAINS.

From the bathroom below pass the end of the remaining length of cable up through the other hole in the ceiling, along and down through the hole at the switch position alongside the other cable.

Fixing the cord-operated switch

Ceiling-mounted double-pole switches for use with showers generally have a mechanical flag as well as a neon. Because of the size of cable, use a 45mm deep surface-mounting box to install them.

Knock out a section of thin plastic in the base of the 1-gang box. Thread in the two sheathed cables and fix the box to the ceiling using No 8 wood screws. Strip about 100mm of sheathing from the end of each cable and about 10mm from the ends of the four insulated wires. Connect the red wire of the cable marked MAINS to the 'L' terminal and the black wire of this cable to the 'N' terminal of the two marked SUPPLY in the ceiling switch. Connect the red and the black wires of the other cable to the 'L' and the 'N' terminals marked LOAD. These terminals may contain two yellow-sleeved wires connected to the neon indicator.

Slip green/yellow PVC sleeving over the two bare earth wires and connect these to the 'E' terminal of the switch. Place the wires in the box and fix the switch with the two screws supplied. Placing 6.0mm^2 cable in the box behind the switch is difficult as it is stiff; it will be slightly easier if the wires are at least 100mm long.

Wiring to the shower unit

The connections at the shower unit differ with each make and model, so it is necessary to follow the installation instructions supplied. All plumbing should be completed before electrical work and cross bonding connections should be taken from the shower's earth terminal to all exposed metal in the bathroom – see page 186. When the shower unit is wired up, turn off the electricity and connect the circuit cable to the consumer unit.

11: Immersion heaters and storage heaters

As a fuel for home heating and water heating, electricity is relatively expensive if used at the full daytime rate. But if the electricity can be used at night under the Economy 7 tariff when the cost is much less (see page 191 for details), electricity is competitive on price with oil, gas and solid fuel.

To heat a whole house electrically, electricity is used at night to charge up *storage heaters* which can then give out their heat during the day. To provide hot water, an *immersion heater* heats up the contents of the hot water cylinder during the night so that the water is ready for use during the day.

Immersion heaters

Electric immersion heaters are fitted into domestic hot water storage cylinders either as the sole means of providing hot water for the sink, bath and wash basins or to provide hot water during the summer, when it could be wasteful to run a central heating boiler. An immersion heater is a heated element (not unlike a kettle element) which is screwed into a boss in the top or the side of the hot water cylinder. A separate thermostat fitted down the centre of the element, controls the temperature of the water by turning the heater on and off. There are three choices of immersion heater, depending on the design of cylinder:
– a *single-element* immersion heater, top or side entry;
– a *dual-element* immersion heater (one short element and one long element), top entry;
– *two elements*, side entry.

With a single element, the whole contents of the cylinder are heated whenever the heater is on; with a dual-element or two elements, there is a choice of heating the whole cylinder (with the longer element of a dual-element heater or the lower of two side-entry heaters) or of heating just the top part (with the shorter element of a dual-element heater or the upper of two side-entry elements). Only the water above the element is heated.

Hot water cylinders

There are two main types of hot water cylinder – direct and indirect.

A *direct* cylinder can be of one of two types. If used only with an immersion heater, it has only two tappings for water pipes – for cold water in and for hot water (and the safety open vent pipe) out. If used with a gas circulator or a back-boiler, it has two additional tappings for taking the contents of the cylinder to the circulator or back-boiler to be heated up. In a hard water area, this latter type of cylinder tends to suffer badly from scale and is best replaced. A special type of direct cylinder is the *Economy 7* cylinder, which is larger than a normal cylinder (so that the hot water available in the morning lasts all day) and is usually built to a higher specification. It has two side-entry immersion heater bosses.

An *indirect* cylinder also has two extra tappings for taking water to and from a boiler, but these are connected to a coil of pipe inside the cylinder such that the central heating water is kept separate from the domestic hot water, thus avoiding the problems of scale (and allowing a corrosion inhibitor to be used in the central heating system).

If you want to use an immersion heater with a cylinder which is not already fitted with a boss, it may be possible to add a boss by means of an 'Essex flange' fitted into a hole in the wall of the cylinder, but generally it will be better to fit a new pre-insulated cylinder.

Immersion heater control

The simplest, but least convenient, way of controlling an immersion heater is to turn it on and off at its wall switch. Even if the house is not converted to Economy 7 (which it should be if the immersion heater is used for anything more than a back-up to a central heating system), it is cheaper to turn off the immersion heater when it is not required, but

152 Immersion heaters and storage heaters

a manual switch means at least two on/off switchings a day.

A better method of control for a single-element immersion heater is to use an *immersion heater timeswitch*, wired between the switch and the immersion heater (see Photo 97), which can be set to bring the heater on when required – or only during the Economy 7 period. The timeswitch can be overriden manually when necessary if the hot water runs out.

A dual-element immersion heater (or two side-entry heaters) can be manually controlled with a *dual switch*: one switch is ON/OFF; the second, marked 'SINK/BATH', selects one element or the other. For use with the Economy 7 tariff, an *Economy 7 controller* replaces the immersion heater wall switch and will automatically bring on the lower of two side-entry immersion heaters or the longer element of a dual element heater during the cheap night-time period. For 'topping up' during the day, a run-back timer on the controller can be set to give a one or two hour boost using the upper (or shorter) element. Economy 7 controllers can also be used with single element heaters, but the whole cylinder will be heated during the boost period.

Installing an immersion heater

You will need to choose the type of immersion heater which suits your hot water cylinder. Top-entry immersion heaters are typically 685mm long with a 457mm thermostat, whilst side-entry immersion heaters are typically 355mm long with a 280mm thermostat. All immersion heaters are rated at 3kW.

A 3kW immersion heater requires its own 20A circuit, run in 2.5mm² cable all the way from the consumer unit (or switchfuse unit). Dual-element and twin-element heaters need the same circuit as only one heater is on at a time.

The materials required for the circuit are: a length of 2.5mm² 2-core and earth flat PVC-sheathed cable; a short length of green/yellow PVC sleeving; a length of 1.25mm² 3-core circular sheathed heat-resisting flexible cord; a 20A double-pole switch (or 20A dual switch or Economy 7 controller); a 1-gang mounting box for the switch (Economy 7 controllers have their own mounting box); a 20A fuseway complete with fuse unit or MCB (or separate switchfuse unit); plastic cable clips, PTFE tape, screws and wallplugs.

Fitting the immersion heater

An immersion heater comes with a large sealing washer, but before fitting it, wrap some PTFE tape round the threads in a clockwise direction. Screw the heater firmly into the boss, tightening it with a special immersion heater spanner.

Wiring the circuit

Fix the mounting box of the control switch on the wall of the airing cupboard within reasonable distance from the immersion heater. From this box run the circuit cable to the consumer unit using the easiest route, which will normally mean raising one floorboard on the landing and drilling holes in the joists to accommodate the cable. Allow sufficient cable for connection in the consumer unit or in a switchfuse unit if there is no spare fuseway. It is necessary that the circuit be for the exclusive use of the immersion heater and that it is not a spur branching off a ring circuit.

Connecting a 20A double-pole switch

For a single-element immersion heater, thread the circuit cable into the mounting box. Prepare the end of the cable by removing about 100mm of outer sheathing, and some 10mm of insulation from the ends of the red and black wires. Slip green/yellow PVC sleeving over the end of the bare earth wire and connect this to the 'E' terminal of the switch. Connect the two insulated wires to the terminals marked either SUPPLY or MAINS.

Remove about 100mm from the sheathing of one end of the 3-core heat-resisting flexible cord and about 10mm of insulation from the end of each flex wire. Connect the green/yellow earth wire to the 'E' terminal. The brown flex wire is connected to the 'L' LOAD terminal of the switch and the blue to the 'N' LOAD terminal.

Place the wires in the box and fix the switch (neon at the top) to the box using the screws supplied with the switch. Cut the

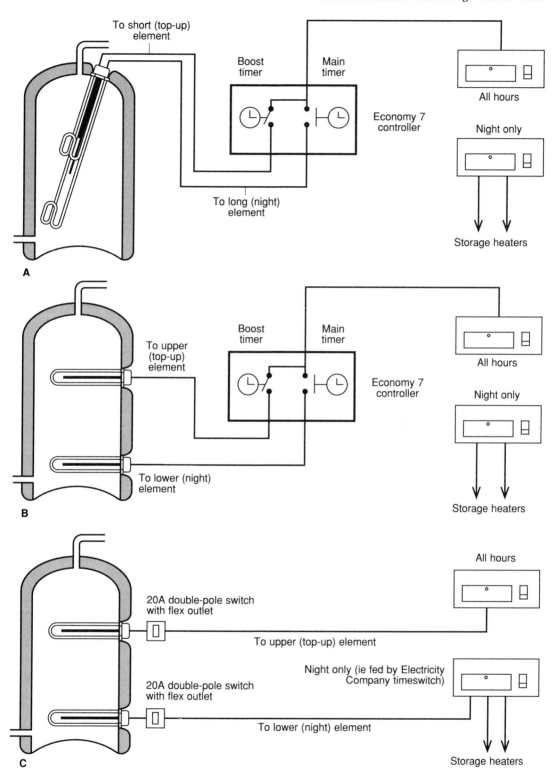

Fig 80 Immersion heater circuit arrangements: **A** Dual-element immersion heater wired via an Economy 7 controller from the main consumer unit; **B** Two side-entry immersion heaters wired via an Economy 7 controller; **C** Two side-entry immersion heaters wired separately – one from main consumer unit and one from the night-time (Economy 7) consumer unit.

154 Immersion heaters and storage heaters

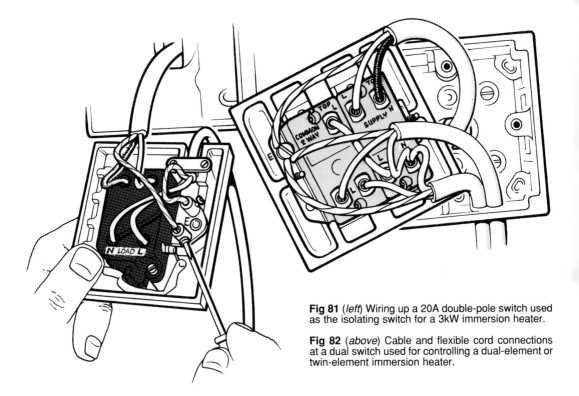

Fig 81 (*left*) Wiring up a 20A double-pole switch used as the isolating switch for a 3kW immersion heater.

Fig 82 (*above*) Cable and flexible cord connections at a dual switch used for controlling a dual-element or twin-element immersion heater.

flexible cord as necessary to reach the immersion heater terminal box and fit the switch to its mounting box.

Connecting a dual switch

Where the immersion heater is of the dual type or there are two side-entry immersion heaters, a dual switch (see Fig 82) is fitted into the switch mounting box instead of a double-pole switch. At the back of the unit it will be seen that there are two separate switches, one being the DP (double-pole), the other a 20A 2-way switch which in this instance is being used as a single-pole change-over switch.

First take a short length of red insulated wire from the 2.5mm² 2-core and earth PVC-sheathed cable. Bare each end; connect one end to the LOAD 'L' terminal of the DP switch and the other end to the 'Common' terminal of the 2-way switch.

Prepare the ends of the circuit cable and connect the red wire to the 'L', MAINS or SUPPLY terminal of the DP switch and the black wire to the corresponding 'N' terminal. Slip green/yellow sleeving over the bare earth wire and connect to the earth terminal.

Measure and cut off two lengths of the 3-core heat-resisting flex, both to reach the heater head of a dual-element immersion heater or, in the case of a side-entry immersion heater, one to reach the upper heater, the other to reach the lower heater.

From one end of each length, strip off about 100mm of outer sheathing, and about 10mm of insulation from each wire. Connect the two blue wires to the LOAD 'N' terminal of the DP switch. Connect the two green/yellow wires to the 'E' terminal of the unit.

Connect the brown wire of one of the flexible cords to the L1 terminal of the 2-way switch and stick a label 'SINK' on the other end of this flex. Connect the brown wire of the other flex to the L2 terminal of the switch. Fix the switch to the box and the cover plate to the grid using the original screws.

Connecting an Economy 7 controller

Prepare the ends of the circuit cable and connect the red and black wires to the LIVE IN and NEUTRAL IN terminals respectively.

Immersion heaters and storage heaters **155**

Photo 95 A 20A double-pole switch with neon indicator for controlling an immersion heater or other electric water heater except a shower unit (*MK*).

Photo 96 A dual switch combining a 20A double-pole switch and a change-over switch for a dual-element or twin-element immersion heater (*MK*).

Slip green/yellow sleeving over the bare earth wire and connect to the earth terminal.

Prepare the ends of two lengths of 3-core heat-resisting flex as for a dual switch, and connect to the off-peak and boost terminals as in the instructions. Connect both earth wires to the earth terminal. Label the cable 'BOOST'.

Connections at the immersion heater

For a single-element immersion heater, strip off about 10mm of insulation from the ends of each of the three wires and thread the flex through the hole in the cover. Connect the brown wire to the empty terminal of the thermostat, the blue wire to the 'N' terminal and the green/yellow wire to the earth terminal. Check that the short wire connected to the thermostat is connected to the 'L' terminal of the immersion heater.

For a dual-element immersion heater, connect the brown wire of the SINK or BOOST flex to the thermostat for the shorter element and the brown wire of the other flex to the thermostat of the longer element. Connect the blue wires to their respective 'N' terminals (or to the single 'N' terminal) and both green/yellow wires to the earth terminal.

For twin-element heaters, connect the SINK or BOOST flex to the upper immersion heater and the other flex to the lower heater.

Replace the immersion heater cover(s) and go to the consumer unit or switchfuse unit to connect up the circuit cable.

Storage heaters

Electric heaters, which make use of the Economy 7 tariff, are made in three principal types: (i) storage heaters; (ii) fan-assisted storage heaters and (iii) combination storage heaters and convector heaters. These are discussed in turn

Storage heaters

Electric storage heaters have an insulated core which can be electrically heated at night and which can then give out its heat slowly during the day. Modern storage heaters are much slimmer and more attractive than the monsters of years ago and are fitted with both input controls and output controls – to regulate the amount of heat taken in at night and to adjust the rate at which it is given out during the day, respectively.

156 Immersion heaters and storage heaters

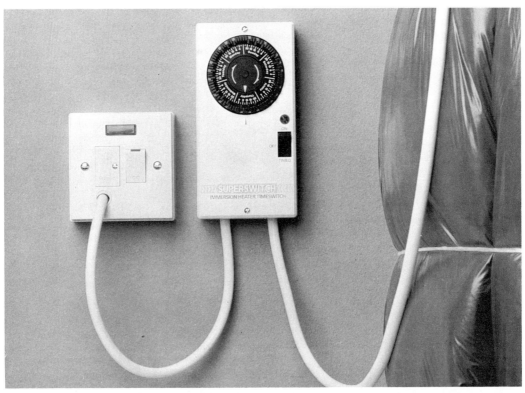

Photo 97 An immersion heater timeswitch is wired between isolating fused connection unit (or double-pole switch) and the immersion heater (*Superswitch*).

Photo 98 *(below)* Modern slimline storage heater (*Creda*).

Immersion heaters and storage heaters

Fan-assisted storage heaters

These operate in the same way as normal storage heaters, but have in addition an electric fan (connected to the normal socket outlet circuit) to boost the output from the heater when required. The fan draws in air from the room and drives it across the heated core sending it back into the room much warmer. The operation of the fan is controlled by an internal thermostat.

Combination storage/convector heaters

One part of this heater works like a normal storage heater, operating on cheap-rate electricity. The other part, wired to a socket outlet circuit, provides heat at full-price electricity via an electric convector heater, thus allowing heat to be provided when the storage heater has run out of stored heat or cannot give a great enough output.

Choice of size

Storage heaters are made in three main sizes with loadings of 1.7kW, 2.55kW and 3.4kW (there is also a smaller size of 0.85kW). Fan-assisted storage heaters generally come in larger sizes – from 2.55kW up to 6.0kW loading.

These loadings represent the *input* rating – that is, the rate at which electricity is taken in at night. The maximum *output* loadings are roughly 40 per cent of these – around 1kW for a 2.55kW heater, for example.

Combined storage/convector heaters have the same ratings for the storage side plus a convected heat output of approximately 1kW, 1.5kW or 2kW for the three main sizes of heater.

Storage heater manufacturers provide instructions on the performance and choice of heater and where it should be positioned in a room. Consult these instructions carefully before choosing and before installing the heaters.

Circuit wiring

Each electric storage heater requires its own separate circuit, except possibly the smallest size where two heaters can be supplied from one circuit.

The reason for separate circuits is that unlike portable heaters and other conventional electric heaters which are switched on and off at any time during a period of 24 hours so producing a diversity of use and reduced maximum current demand on the circuit, usually a multi-outlet ring or radial circuit, storage heaters are all switched on at the same time and all are consuming current during the short off-peak period overnight. This means there is no diversity so a ring circuit is out of the question for storage heaters.

Installing storage heaters

The circuit for a storage heater, and for the storage heater of a fan-assisted storage heater or a combined storage heater/convector heater, is a radial circuit, run in 2.5mm² 2-core and earth PVC-sheathed cable, supplied from a 20A fuseway in the Economy 7 night-time consumer unit, fitted with a 20A fuse or miniature circuit breaker (MCB).

The consumer unit is of conventional pattern having one way for each heater circuit plus any spares, one of which can be used for an immersion heater circuit.

Storage heaters

For a simple storage heater, the connection from the circuit to the heater is via a 20A double-pole (DP) switch, with a flex outlet and, preferably, fitted with a neon indicator. The final connection is with 1.5mm² heat-resisting flexible cord, wired to the heater as described in the installation instructions supplied with the unit.

Fan-assisted storage heaters

Although the heater of the fan-assisted storage heater is connected to a 20A circuit from the storage-heater consumer unit, a circuit is also required for the fan. As the fan is switched on at any time during the 24 hours, its circuit must be connected to the unrestricted 24-hour electricity supply and not the Economy 7 circuit.

The best way to provide this connection is to run a spur from the ring circuit. A fused connection unit (with a 3A fuse) is inserted into the spur and from the connection unit a 1.5mm² 2-core and earth cable is run to the storage-heater mains outlet.

158 Immersion heaters and storage heaters

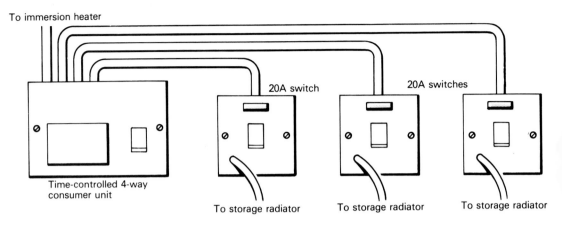

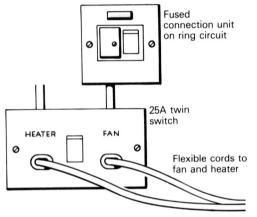

Fig 83 *Top*: Circuit wiring for three storage radiators plus an immersion heater. *Left*: Twin 25A isolating switch for a fan-assisted storage heater. The supply for the fan comes from a fused connection unit on the socket outlet circuit.

Photo 99 A 25A twin double-pole switch for use with fan-assisted storage heaters (*MK*).

As there are two circuits with two flexible cords running into the heater – one for the heater, the other for the fan – a special twin switch has to be fitted at the mains outlet instead of a double-pole 20A conventional plateswitch (see Fig 83). The twin switch is of 25A rating and comprises two DP switches on the one faceplate controlled by a single rocker so that both switches are switched off simultaneously to isolate the heater and fan from the mains supply. The twin switch has two flexible cord outlets in the faceplate marked HEATER and FAN respectively.

The circuit wires are connected to the SUPPLY terminals of the respective switches and the flexible cords to the LOAD terminals.

Combined storage/convector heater

The storage heater part of a combined appliance is connected in the same way as a normal storage heater – that is to a 20A double-pole switch wired to the night-time consumer unit.

The convector part needs a connection to the 24-hour supply – either via a plug into a convenient socket outlet or, for a more permanent installation, via a fused connection unit or double-pole switch wired as a spur from the ring circuit.

Photo 100 A 20A double-pole switch suitable for providing the supply to a storage heater. Versions without neon and with front flex entry also available (*MK*).

Single storage heater

A single storage heater, perhaps in a conservatory or loft conversion, does not need to be wired to a separate consumer unit. Provided it has an input rating of less than 3kW, it can be connected to the local socket outlet circuit via a timeswitch to bring it on during the Economy 7 period.

12: Electricity outside the house

It is extremely useful to have electricity outside the house. Not only in the garden for powering garden equipment (and perhaps lighting), but also in the greenhouse for heaters and plant propagators and in the garage and/or workshop for power tools, heating and lighting.

To get electricity to a greenhouse or a garage or even to a shed used as a workshop means running a fixed cable from the house. To use instead a flexible cord plugged into a house socket outlet and slung between the house and, say, the shed is unsatisfactory, is potentially dangerous and has caused accidents, some fatal.

The power supply to an outside building must be run as a separate circuit from the consumer unit or fuse box and must be protected by a residual current device (RCD) – see page 187.

There are two approved methods of installing the section of outdoor cable between the house and the detached 'building'. One method is to run the cable overhead, the other is to bury it in the ground.

Overhead wiring

Ordinary 2-core and earth flat PVC-sheathed house wiring cable may be run overhead out of doors. Where the length of span does not exceed three metres no intermediate supports are required for the cable, provided that where the cable is supported at each end no undue strain is placed on the conductors and there is no likelihood of the cable sheathing being chafed. Where the length of span exceeds three metres it is necessary to support the electric cable by attaching it to or suspending it from a catenary wire fixed to the house and to the outbuilding. The catenary wire can be a stranded galvanised wire, which is available from electrical wholesalers. Even for spans shorter than three metres, using a catenary wire is better than having an unsupported length of electric cable. The catenary wire itself should be earthed with a length of 2.5mm² single-core cable taken to the main earthing point.

Although it is convenient and fairly simple to install an overhead cable for short spans with a catenary, it is a totally different proposition to install an overhead cable on longer spans. Apart from it being an eyesore there is always a risk of the cable being blown down in a gale, with serious consequences. Also for very long spans it is necessary to fix poles to support the cable and to prevent undue sagging.

An alternative for spans of less than three metres is an unjointed length of heavy-gauge galvanised steel conduit, firmly fixed, earthed and provided with PVC bushes at the ends.

There are three ways to provide the necessary RCD protection. One is to connect the circuit to a spare MCB on the consumer unit which is already RCD protected. A second is to use an existing 'way' on the consumer unit (not protected by an RCD) and then to fit an RCD in the cable before it passes outside (a method which could also be used with a new switchfuse unit if there are no spare fuseways on the consumer unit). The third is to use a combined RCD/MCB unit as the new switchfuse unit which will provide both the extra circuit required and give it residual current device protection.

Materials required

The materials required for an overhead span of cable starting at or near the consumer unit and running to a switchfuse unit or consumer unit in the garage, greenhouse or shed are: a length of 2-core and earth 2.5mm² flat PVC-sheathed cable for a 20A radial circuit (4.0mm² for a 30A radial circuit); a length of stranded galvanised cable, for the catenary wire; two eye bolts; a wire tensioner; cable fasteners; an RCD unit for the house end of the circuit if required; a new switchfuse unit (or combined MCB/RCD unit) for the house end of the circuit if required; a splitter unit if a new switchfuse or RCD/MCB unit is used; a switchfuse unit or 2-way consumer unit for the other end of the cable run in the outbuilding; a length of green-yellow PVC-insulated earth cable and a length of green/yellow PVC sleeving; plastic cable clips and wood screws.

Installing an overhead cable

An overhead cable must be fixed at a height of not less than 3.5m above general ground level or 5.2m above a driveway.

Fix the eye bolts at least 300mm higher than the minimum permitted height to allow for sagging. Fit the cable tensioner to one eyebolt and join the catenary wire to each. If the regulation minimum height cannot be obtained at the garage, greenhouse or workshop end fix a pole or piece of timber at this point.

Drill a hole in the outside wall of the house and another under the eaves of the garage. Attach an earth clamp to the catenary wire and connect the earthing cable to it. Then thread the earth cable and the circuit cable through the hole in the house wall and take it to the position of the consumer unit or switchfuse unit (if necessary connecting it through an RCD unit) or combined MCB/RCD unit. In the outbuilding, take the cable to the position of the new switchfuse unit or 2-way consumer unit and connect it up. Turn off the electricity before connecting to the consumer unit, switchfuse unit or combined MCB/RCD unit; if a splitter unit is required, make the connection to this and then ask the electricity company to come and connect the unit up to the meter.

Underground wiring

Burying in the ground is the better method of running a cable from the house to a greenhouse, detached garage or workshop. Once laid the cable is out of sight, unlikely to be disturbed and will give good service over a long period of years.

The type of cable to use for the underground part of the circuit is armoured cable. This has three PVC-insulated wires inside a PVC sheath, which is further covered with a steel armour and a PVC cover. Provided it is buried deep enough in the ground not to be disturbed by digging (which means at least 500mm deep under flower beds or vegetable

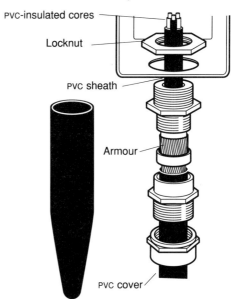

Fig 84 3-core armoured cable plus compression gland used for wiring outdoor circuit.

plots), no further protection is needed.

Where it enters metal mounting boxes, armoured cable needs special compression screwed glands (see Fig 84).

Only the outdoor section of the cable need be of this special and costly type. In both the house and the outbuilding the cable can be terminated at a junction box. From this box ordinary PVC-sheathed house wiring cable is used, the two types of cable being jointed in the junction boxes, which can simply be surface-mounting boxes.

An alternative to using armoured cable is to use normal PVC-sheathed and PVC-insulated 2-core and earth cable protected by rigid PVC conduit underground. No junction boxes are required and elbows and couplers available for the PVC conduit can be used to take the cable into the house and into the outbuilding.

Materials required

The materials required are: a length of 3-core PVC-covered wire armoured cable of the appropriate size and current rating (2.5mm^2 for a 20A radial circuit; 4.0mm^2 for a 30A radial circuit); two compression screwed glands, one for each end of the cable, two 1-gang knock-out metal boxes, together with two 3-way terminal connector blocks and box covers and two semi-blind PVC grom-

162 Electricity outside the house

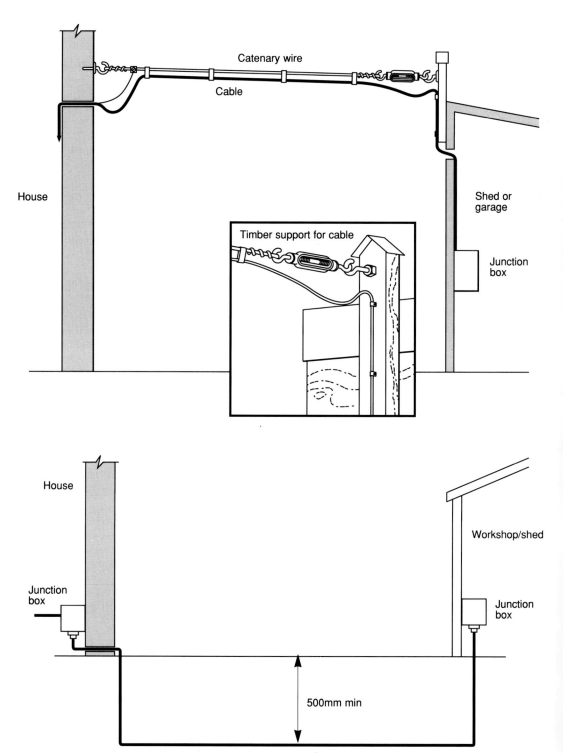

Fig 85 The outdoor section of a cable running between the house and a detached garage, shed or greenhouse, showing the overhead and underground cable of the alternative methods.

mets to fit the knockout holes; a length of 2-core and earth flat PVC-sheathed cable of the same current rating as the outdoor section of cable; an RCD unit for the house end of the circuit if required; a new switchfuse unit (or combined MCB/RCD unit) for the house end of the circuit if required; a splitter unit if a new switchfuse or RCD/MCB unit is used; a switchfuse unit or 2-way consumer unit for the other end of the cable run in the outbuilding; rigid PVC conduit and fittings plus PVC-sheathed cable as an alternative to armoured cable; green/yellow sleeving; plastic cable clips; woodscrews and wallplugs.

Installing the outdoor section of cable

Decide where the cable is to pass into the house at a point above ground level and above the damp course but if possible under the floorboards. Decide also at what point the cable is to enter the greenhouse or garage. Cut a hole in the outside wall of the house for the cable and drill a hole in the outbuilding.

Plan the route for the trench avoiding vegetable and flower beds, preferably keeping to the edge of the lawn. Dig the trench to a depth of 500mm, or deeper. Remove any flints or sharp stones and place about 50mm of sand or sifted soil in the bottom of the trench. Lay the cable and pass the ends through into the house and outbuilding. Fill in the trench removing any flints from the soil.

Indoor section of cable

Knock out two blanks from the metal knockout box. Fix the box to timber under the floorboards in the house. Strip off about 250mm of outer PVC covering and wire armour from the end of the armoured cable. Fit the compression gland. Thread the ends of the wires into one knockout hole, screw a backnut on to the gland and tighten it to provide a good metal-to-metal electrical connection. Fit a PVC grommet into the other knockout hole and thread in the end of the PVC-sheathed cable after first piercing a small hole in the grommet. Prepare the ends of the PVC-sheathed cable and join the two sets of wires in the terminal block: red wires in one outer terminal, black wires in the other outer terminal and earth wires in the centre terminal (cover the bare earth wire with green/yellow sleeving). Fit the cover of the box and repeat the procedure for the junction box in the outbuilding.

Connecting the circuit to the consumer unit (or to a switchfuse unit or combined RCD/MCB unit) is described on pages 182-4.

Lighting and power in outbuildings

The wiring of lighting and power points in the home garage and workshop is carried out in ordinary PVC-sheathed cable. Switches, socket outlets and light fittings may be moulded plastic as used in the house, or for greater mechanical strength metal-clad switches and socket outlets are preferred. Flexible cord pendants should not be fitted. For maximum light output at least one fluorescent fitting is recommended, but do not rely solely on fluorescent lighting if there is moving machinery. For additional lighting, batten lampholders and adjustable lights should be fitted.

In the outbuilding, lighting and power can be on separate circuits provided from a 2-way consumer unit – a 20A circuit for power and a 5A circuit for lights – or the lighting can be run from a fused connection unit in a single circuit wired into a switchfuse unit. If no RCD protection is provided inside the house (see above), an RCD must be fitted in the outbuilding to protect the whole circuit or the socket outlets themselves must be RCD protected – see page 187.

For additional protection inside a garage, greenhouse or workshop, the cable between sockets, light fittings, switches and the like can be run in PVC conduit.

Installing outdoor socket outlets

Socket outlets situated outside the house are convenient for the electric mower, hedgetrimmer, power tools and other tools and portable apparatus used out-of-doors. Without one or more outdoor socket outlets, mowers and hedgetrimmers and other tools are run off house sockets which is not only inconvenient but can cause accidents.

Position of socket outlets

Probably the best position for an outdoor socket outlet is on the outside wall of the house. When the socket is fixed in this posi-

164 Electricity outside the house

Photo 101 Metal-clad switched fused connection unit with neon indicator especially suitable for the workshop and garage (*MK*).

Photo 102 Metal-clad switched socket outlet, designed for use with non-standard 13A plugs (*MK*).

tion, the cable supplying the socket can pass through the outside wall at the back of the socket into the socket box without any cable running outside the house. This is a particularly convenient way of wiring the socket if it is being taken as a spur from an existing socket on the *inside* of the outside wall – see *Wiring the circuit*.

Carefully choose the position for the socket outlet(s), first checking that it will be possible to cut a hole through the wall immediately behind the socket. A sink, furniture or even decorations may prevent a hole being cut in some positions.

Cut the hole before running the circuit cable. The best method is to use an electric drill fitted with a long masonry bit. Angle the drill slightly downwards to the outside, to prevent moisture getting in, and fit a short length of PVC conduit into the wall to protect the cable.

Outside, the socket outlet should be fixed about 600mm above ground level which in the house brings the hole in the wall above skirting level. If drilling through a cavity wall, it may be possible to offset the holes in the inner and outer leaf, passing the cable up the cavity.

Wiring the circuit

The circuit for an outdoor socket outlet can conveniently be wired as a spur branching off the ring circuit at the terminals of a socket outlet in the house or from a junction box inserted in the ring cable under the floor of the house.

A separate spur is required for each outdoor socket outlet. All outdoor sockets (and, in fact, any sockets supplying equipment used outside) must be RCD protected. This means either that they are connected to a circuit which is RCD protected, or that an RCD is inserted in the cable leading to the outside socket or that the socket itself is RCD protected.

Electricity outside the house 165

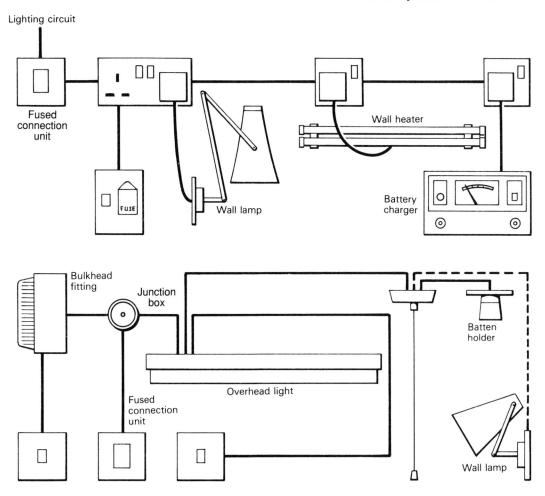

Fig 86 *Top*: Radial power circuit supplying 13A socket outlets for the equipment used in the home workshop, and a fused connection unit to supply the workshop lighting. *Bottom*: Layout of workshop lighting circuit, light fittings and switches.

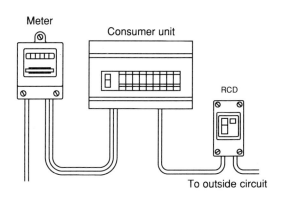
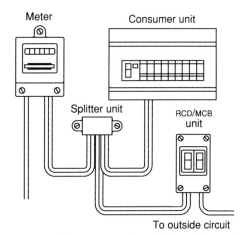

Fig 87 Internal wiring for a garage/workshop circuit. *Left*: Where the consumer unit does not have RCD protection, fit an RCD in the circuit; *Right*: Where the consumer unit has no spare 'ways', fit a splitter unit plus a combined RCD/MCB unit.

166 Electricity outside the house

Photo 103 'Weatherseal' exterior socket outlet (*Crabtree*).

Wiring and fixing an outdoor socket outlet

An ordinary moulded plastic socket outlet may be installed out of doors provided it is protected from the rain and from the risk of breakage. The socket outlet should therefore be protected by a small enclosure of timber fixed to the outside wall. In addition a splashproof cover can be fitted over the socket. In some situations the latter may be all that is needed to afford protection.

A far better choice, however, is a purpose-built exterior socket, which comes with its own enclosure mounting with a gasket sealing the socket outlet to the mounting box. The mounting box will accept conduit fittings and the socket itself is fitted with a snap-shut transparent flap cover. Versions are available with RCD protection built in and with ON/OFF switches. Special outdoor plugs are also available which are specifically designed to fit into this outdoor socket outlet.

Connecting up this type of socket is exactly the same as for an ordinary 13A socket outlet.

Connections to the ring circuit

Turn off the power before attempting the connection. Whether the connection of the spur cable is made at a socket outlet or at a junction box inserted in the ring cable, the procedure is the same.

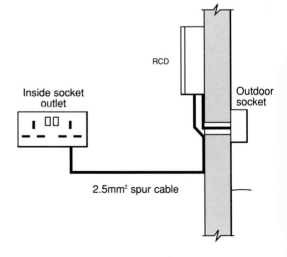

Fig 88 Outdoor socket outlet wired as a spur from a socket outlet circuit with a high-sensitivity RCD protecting against electric shock.

Electricity outside the house 167

Photo 104 'Weatherseal' exterior socket outlets. *Left*: Switch/socket combination; *Right*: single socket; *Inset*: Sleeved plug (*Crabtree*).

Installing garden socket outlets

Socket outlets are useful in the garden, not only for the mower and hedgetrimmer in a large garden, but for garden lighting and pool lighting with miniature fountain and waterfall power supplied from a mains transformer which can be plugged into a garden socket outlet.

Choice of socket outlet

Because of the risk of damage, ordinary plastic moulded socket outlets should not be used further down the garden, even if in an enclosure. Metal-clad versions are acceptable, however, provided they are mounted in a weatherproof enclosure, such as a wooden box, made from durable timber.

The exterior sockets described earlier and shown in Photos 103 and 104 can be used in the garden, provided they can be mounted on a firm vertical surface. A more expensive alternative is to use heavy-duty metal exterior socket outlets which have a watertight cover which screws firmly over the socket outlet when the socket is not being used.

Installing the circuit cables

Provided rigid PVC conduit is used throughout, normal PVC-insulated PVC-sheathed cable can be used for supplying outdoor socket outlets: the conduit is taken right up to the socket mounting boxes and joined with special connectors so that the whole circuit is weatherproof.

Alternatively, armoured cable can be used to provide the supply to the sockets, but this should be well protected where it comes out of the ground to go up to a socket outlet.

The circuit in PVC cable or armoured cable is run in the same way as described earlier for a circuit to an outbuilding – that is via a

168 Electricity outside the house

junction box inside the house (not necessary with PVC cable) back to the consumer unit or RCD/MCB unit. A separate RCD must be fitted if the circuit in the consumer unit is not RCD-protected or if the supply is from a switch-fuse unit. Wherever possible, a double-pole switch should be built into the circuit, so that the garden circuit can be turned off when it is not being used.

Some electricians will use mineral-insulated copper-clad (MICC) cable for running outdoor wiring, but this needs special seals fitted on to the ends which is not a do-it-yourself job.

Garden lighting

For temporary lighting of a garden area, such as a patio, mains-voltage 'festoon' lights (a series of coloured lamps in a string) can simply be plugged into a convenient socket outlet indoors and then unplugged at the end of the evening.

Mains-voltage lights on the house wall have been discussed earlier (in Chapter 5) and mains-voltage lights can be installed permanently in the garden, wiring the circuit as described for a garden socket outlet. But the easiest type of garden lighting to install is *low-voltage* lighting.

Photo 105 Low-voltage feature garden spotlight with spike. Wall-bracket and choice of coloured lenses available (*Hozelock*).

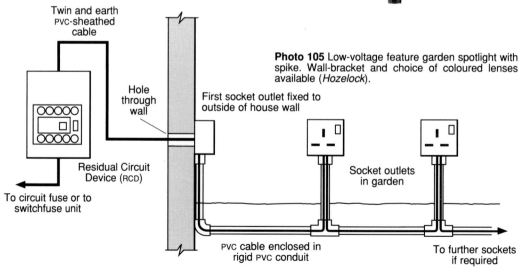

Fig 89 Socket outlets situated in the garden, supplied from the house ring circuit and protected by an RCD.

Electricity outside the house 169

Photo 106 Low-voltage garden feature spotlights, showing version on spike and version on wall bracket, enhancing the garden at night (*Hozelock*).

This comes with a mains-fed transformer, which you can plug into a socket outlet inside the house – or, if more convenient, in the shed, garage or greenhouse – and simply run the low-voltage cable down the garden along the surface of the soil as there is no danger of an electric shock.

Transformers have an output of a safe 24V and one transformer will operate several outdoor lights – typically up to six lights in a single circuit.

The lights themselves come in various shapes and sizes, the most popular of which are plain or coloured spotlights on a spike which pushes into the ground and spotlights on a bracket which can be attached to the wall. Other options include globe lights, shaded lights (with a cover on top so the light is directed downwards), tiered lights and cylindrical lights.

13: Rewiring a house

Some houses may still have old wiring which has reached the end of its useful life and is probably posing a danger of fire if not of electric shock. If your house is like this, rewiring is a must.

But you may be thinking of partially or wholly rewiring simply because your home doe not have an adequate number of circuits, socket outlets or lighting points. For years, many builders have been putting in the bare minimum number of socket outlets, which may have been adequate in the 1950s or 1960s, but are ludicrous in the 1990s, with the enormous growth in domestic appliances, and a desire to have more lighting points in rooms than the single central pendant light fitting which is generally supplied.

Rewiring, whether whole or partial, is a major undertaking and not one to be undertaken lightly. It is a job done best in the summer, when there are more hours of daylight and no need for heating, since the electricity is likely to be off for considerable periods of time.

Planning rewiring

There are several things to think about when rewiring a house: the number of circuits; how and where to provide RCD protection; the number of socket outlets; the type and number of lighting points; and what other circuits you need.

Number of circuits

Most houses will have six circuits – two socket outlet circuits, two lighting circuits, a cooker circuit and an immersion heater circuit. But you might want to think about having an additional socket outlet circuit for the kitchen, a circuit for a shower, a circuit for outdoors, plus perhaps a spare for future extension – making a total of ten.

RCD protection

It is essential that all socket outlets which might be used to supply equipment used outdoors are RCD protected. But if you are rewiring, it is a good opportunity to provide RCD protection for all socket outlets, either by fitting an RCD to each socket outlet circuit (if you are keeping the existing consumer unit) or by having a consumer unit with a built-in RCD – see Chapter 14 for details.

Number of socket outlets

The table on page 177 gives recommendations for the number of socket outlets in the average home. But even these should be regarded as a minimum. Modern wiring is expected to last for at least 30 years, so the wiring of a home should be installed with that in mind and there is no doubt that the number of electrical gadgets in a house will increase. Socket outlets themselves are relatively inexpensive and it is much easier to put them all in in one go, rather than face the disruption caused by trying to add extra socket outlets later – to say nothing of the time and expense of redecorating. Also bear in mind that you may want to rearrange furniture, which could mean sockets in different places.

Lighting points

Rewiring is a good opportunity to consider the number and type of lighting points which you might want. In particular, do you want wall lights as well as (or instead of) central lighting points? Do you want a 2A lighting circuit (see page 175)? And do you want to fit Luminaire Supporting Couplers (LSCs) or 'Klik' fittings (see page 43) from the word go?

Inspecting an installation

Where wiring is suspect, the electricity company or an approved electrical contractor should be called in to inspect, test and submit a report on the condition of the various circuits, stating what is required to bring the installation up to standard and what rewiring if any is necessary. It is recommended that such an inspection and test is carried out at intervals not exceeding five years even where the installation is not suspect.

There is however much that the householder can do to satisfy himself that the wiring is in good order or whether he should proceed with rewiring. Some simple tests can also be made but expensive instruments are required for the more important ones.

The parts of the wiring which should be checked are: the main switchgear, ie the consumer unit and/or other main switches and fuses; circuit cables *in situ*; ends of cables within wiring accessories (mainly light switches and socket outlets); and earthing arrangements.

Main switchgear

If the sole main switch and fuse gear is a modern consumer unit, usually enclosed in a plastic casing, check that the casing is not cracked or broken with pieces missing, that the fuse cover is not missing and that no fixing screws are missing.

Turn off the mainswitch and remove the fuse cover. Check that none of the fuse holders is broken and none is missing leaving a fuse wire attached to the fuseway terminals and exposed to touch. Check the current ratings of the fuses against the circuits they protect and, if rewirable fuses are used, check that the fuse wires are of the correct sizes for the fuse in accordance with the fuse colour code (see page 17). Check that the unit is securely fixed to the wall.

Check that the cables connecting the unit to the meter are in good condition. Check the circuit cables, which in a modern installation will have a grey or white PVC sheath. No bare wire must be showing.

If the circuit cables appear to be old and are not PVC-sheathed the indications are that the cables are old but a modern consumer unit has been installed in recent years without the installation being rewired.

If instead of a single consumer unit there is a number of old-fashioned mainswitch and fuse units, some with two fuses, others with more than two fuses and possibly one having a single fuse, these should be scrapped without delay and a modern consumer unit fitted and at the same time some or all circuits should be rewired.

Some old mainswitch and fuse units and some fuseboards have what is termed double-pole fuses. Each circuit has two fuses, one in the live pole, one in the neutral pole. Only single-pole fusing is now permitted. Double-pole fusing is potentially dangerous for should the neutral fuse blow, the circuit remains live although not working. Any such switches or fuseboards should be replaced without delay.

Although a single consumer unit is the normal arrangement for a dwelling and covers all circuits, where night storage heaters are installed these will have a separate consumer unit making two for the installation. Where an electric cooker circuit, a shower unit circuit or any other circuit has been added this is connected to a switchfuse unit installed for the purpose and connected separately to the electricity supply. Therefore the presence of more than one consumer unit or switchfuse unit does not necessarily indicate an old installation

Circuit cables in situ

It is usually only necessary to inspect the cables in the roof space to get a good idea of the condition of the installation and possibly the age. If the cables are neatly run and clipped to the joists and are obviously PVC-sheathed, the wiring can be assumed to be generally satisfactory. Inspect also the ends of the cables where they pass through the ceiling into light fittings. If wires are exposed where the sheathing is removed to connect the wires, these need attention.

If the cables are obviously tough rubber-sheathed, obviously in poor condition, not properly secured with cable clips, junction boxes not fixed properly or covers loose or missing, a rewire is an urgent necessity.

Where all or part of the wiring is in conduit with the ends of conduits cut short above light fittings and not terminating in conduit boxes, this too indicates the need for a rewire.

Similar inspections should be made under floorboards.

Wiring accessories

Old-pattern light switches, ceiling roses and other accessories of a lighting circuit indicate that the installation and therefore the cables are old. Remove a switch from its pattress and check the wires. If the insulation is rubber it will be cracked and hard. Also take down a ceiling rose and make a similar inspection. Where the cables of an installation are of the tough rubber-sheathed type, but most if not all switches and ceiling roses are modern, release a switch from its box and carefully examine the wires for deterioration of the insulation. Remove the ceiling rose cover and make a similar inspection. If the insulation has perished a rewire is needed.

The old-type round-pin socket outlets are almost sure to be connected to old circuit wiring. This should be scrapped and a ring circuit installed.

The modern 13A socket outlet connected to the old circuit cables can be particularly troublesome and possibly dangerous, for these may have replaced old 15A 2-pin socket outlets with no earth conductor installed which means there is no earthing

In many instances a ring circuit will have replaced the old circuits supplying the round-pin sockets but the lighting circuit will not have been rewired.

Earthing arrangements

The importance of effective earthing cannot be over emphasised. Most modern installations will be earthed via an earthing terminal provided with the electricity company's supply cable, but some rural properties may rely on an earthing electrode and some homes may have protective multiple earthing (see page 187 for details).

Apart from visually checking the type of earthing that you have and ensuring that the wires are securely attached, there is little else you can do. An electrician will have the equipment to be able to test the earthing for you and make recommendations.

Wiring lighting circuits

There are two methods of wiring a home lighting circuit: (i) the loop-in method and (ii) the junction-box method, though many lighting circuits are a mixture of both.

Loop-in method

The loop-in method of wiring a lighting circuit consists of running the circuit feed cable containing the live and neutral conductors (as well as the earth continuity conductor) from the circuit fuseway in the consumer unit to each lighting point, ending at the last (see Fig 90). At each lighting point the feed cable is looped into the terminals of the ceiling rose and looped out to go to the next light, and so on up to the last light on the circuit.

The modern ceiling rose, often termed a loop-in ceiling rose, therefore has terminals for the live and neutral conductors (and an earth terminal). The live and neutral terminals are multi-terminal banks or blocks for the various conductors.

From each ceiling rose wired on the loop-in method, another 2-core and earth cable runs to the switch controlling that light. One of these two conductors is a live conductor feeding the switch; the other is the return wire from the switch to the light which is usually termed the switch return wire or the switch wire. The three terminal banks – live, neutral and switch wire – are arranged in-line in most patterns of ceiling rose with the live bank in the centre and the neutral and switch wire terminal banks either side of the live.

The two conductors of the pendant flexible cord are connected to the neutral and switch wire terminal banks. Where 3-core flex is connected to a ceiling rose the earth core is connected to the common earth terminal containing the circuit earth continuity conductors.

Although it is usual to wire all lighting points in a line this is not an invariable rule. One or more of the lights on the circuit can be taken off the in-line circuit in order to save cable or to ease the wiring work. To do that would mean more than three sheathed cables at a ceiling rose. A ceiling rose will readily accommodate a fourth 2-core and earth sheathed cable but not more. It is therefore best to stick to the in-line rule where practicable and leave the introduction of a fourth cable for future extensions (see Fig 27, page 55).

Other light fittings

Although the loop-in system is intended especially for use with the loop-in ceiling rose, it can be and is used where the lighting outlets are other than ceiling roses.

Batten lampholders, which are simple

Rewiring a house

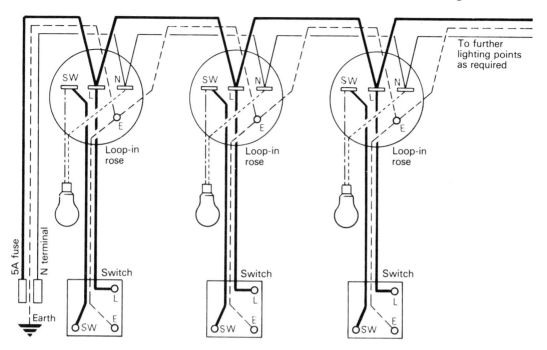

Fig 90 The loop-in ceiling rose method of wiring a lighting circuit.

close-mounted ceiling lights used as an alternative to a pendant, are available in the loop-in pattern. Most other types of light fitting do not have facilities for the live loop-in conductor but the conductors can often be jointed in a cable connector which is housed in the ceiling plate of the light fitting or in a mounting box situated above the ceiling plate of the fitting. Where a cable connector for the live loop-in conductors cannot be accommodated in such a box or in the ceiling plate of the fitting, the circuit cables are terminated in a junction box fixed in the ceiling void and a single 2-core and earth sheathed cable run into the light fitting.

Junction-box method

The junction-box method of wiring a lighting circuit consists of running the 2-core and earth feed cable from the fuseway in the consumer unit to a series of junction boxes which in turn supply the lights and their switches (see Fig 91). The junction boxes are moulded plastic, round in shape, and contain four terminals. Normal wiring practice is to provide one junction box for each light and its switch, situated in a convenient and accessible position for wiring. The junction boxes are fixed to timber in the ceiling void, and where under a boarded floor they should be situated at positions where floorboards can easily be raised so that fixing the boxes and joining the wires presents no problems. In an unboarded roof space junction boxes can be situated anywhere between lighting points and their switches, but under floorboards on the first floor of a conventional house it is usually better to group the junction boxes under a floorboard on the landing. More cable will be required where all or most of the junction boxes are situated under one floorboard, but this method will save lifting other floorboards and the junction boxes will be readily accessible.

From each junction box one length of 2-core and earth sheathed cable is run to the respective light and another length of the cable is run to its switch. This means there is only one cable at each light and four cables at each junction box (see Fig 30, page 59) except the last on the circuit where there are only three cables.

A good-quality junction box having a

174 Rewiring a house

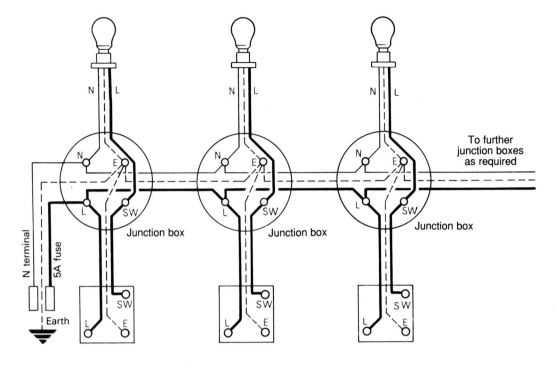

Fig 91 The junction-box method of wiring a lighting circuit.

diameter of 100mm will accommodate more than four cables and therefore can be used for supplying extra lights as required (see Fig 31, page 61).

The four terminals of a junction box are used as (i) live, (ii) neutral, (iii) switch wire, (iv) earth. The terminals are not marked so which is used for which wires does not matter provided the connections are properly done.

There is also a junction box having six terminals. These six-terminal boxes are used where more than one light and switch are supplied from the one junction box, and also for complicated circuits such as 2-way switching when junction boxes are used in the actual switching circuit (see page 98 for details).

Combined loop-in and junction-box circuits

A combined loop-in and junction-box lighting circuit has some lights wired on the loop-in method using loop-in ceiling roses or other light fittings having loop-in facilities, and some lights wired on the junction-box method (see Fig 28, page 58). Which method is used for a specific light usually depends as much on which method is the easier as on the type of light fitting.

For example, where the cable to a light is to be run on the surface of the ceiling in mini-trunking (and, perhaps cornice trunking – see page 37), it is usually better to use a single cable from a junction box, rather than the three cables necessary for loop-in wiring, though larger sizes of mini-trunking will be able to accommodate all three cables.

Porch lights, wall lights, fluorescent lights and many others are better wired on the junction-box method with one cable only running to the light, especially where they do not have loop-in facilities.

Another situation where the junction-box method is preferred is where the switch is a long distance from the light, so that by inserting a junction box, cable can be saved. In general therefore a remotely positioned light should be wired on the junction-box method.

It follows therefore that in most lighting circuits some lights are wired on the loop-in system while others are on the junction-box system.

Choice of method

Apart from the specific situations referred to, the first choice should be the loop-in method. This system has the advantage that the various joints are made at a ceiling rose which is normally accessible, with the joints for the specific light and switch being in the same room. Also a junction box is relatively difficult to install and the wires difficult to connect, representing extra work. When fixed and wired a junction box is inaccessible so should a fault develop or an extension be made to the circuit the junction box has to be located, floor coverings lifted and floorboards raised.

2-amp lighting circuits

Ceiling lights and wall lights will be wired so that they can be turned on and off at light switches by the door. But table and standard lamps, plugged into socket outlets, usually have to be turned on and off one by one. A way of being able to turn all the table and standard lamps in a room on and off at a light switch by the door is to install a 2A lighting circuit.

This utilises 2A round-pin unswitched socket outlets, which were widely used in much older wiring installations, but which are now available in versions to fit into a standard 1-gang mounting box. Each lamp is then wired to a 2A round-pin plug which is plugged permanently into a 2A socket, which will also free the 13A sockets for other uses. No other equipment can be plugged into the 2A sockets, unless it is fitted with one of these special plugs.

To wire a 2A lighting circuit, you can use the junction-box method, with a cable going to the wall switch and then a radial circuit for the 2A socket outlets, with a 1.5mm^2 2-core and earth cable going to the first 2A socket and then on to each 2A socket outlet in turn. The 2A socket outlets themselves should be positioned as close as possible to the lamps they serve, not necessarily at the same height as 13A socket outlets.

Domestic power circuits

A power circuit in the home refers to a circuit supplying 13A socket outlets which accept 13A plugs and supplying fused connection units, fitted with 3A or 13A fuses, which provide power to at least some fixed appliances.

There are two types of circuit used for supplying 13A socket outlets: (i) the ring circuit and (ii) the radial circuit.

Ring circuits

A ring circuit consists of a 2-core and earth PVC-sheathed cable which starts at a 30A fuseway in the consumer unit, runs through the various rooms including landing, hall and other areas, is connected to the socket outlets in those rooms and areas and returns to the same terminals of the fuseway thus completing a loop or ring (see Fig 92). Each ring circuit is limited to supplying socket outlets and fixed appliances over an area of 100m^2 but the number of socket outlets and fused connection units which may be supplied from any one ring circuit is unlimited. Fused connection units may be used with appliances up to 3kW, but water heaters larger than this (and all immersion heaters) need their own special circuit.

Ring circuit spurs

In addition to the ring cable, non-fused spurs may be branched off the ring cable (see Fig 62, page 118). Spurs are intended primarily to supply remotely situated socket outlets off the main route of the ring cable but they are also used to supply fused connection units for special appliances and apparatus such as sink waste disposal units, central heating pumps, wall heaters and other fixed appliances.

Spur cables are connected to the ring cable either at the terminals of a ring socket outlet or at a 30A junction box inserted into the ring cable. The ring circuit spur is a convenient and fairly easy means of adding socket outlets to a ring circuit (see Fig 62).

The number of non-fused spurs connected to a ring circuit is limited to the number of socket outlets (single or double) and fused connection units actually connected to the ring circuit cable. Each non-fused spur may supply one single or one double 13A socket outlet or one fused connection unit.

There is no limit to the number of *fused* spurs connected to the ring circuit. A fused spur is where cable rather than flex is

176 Rewiring a house

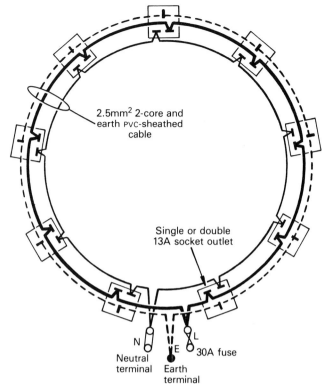

Fig 92 Cable connections of a ring circuit which may supply socket outlets (and fixed appliances via fused connection units) over a maximum floor area of 100m² for each circuit.

connected to a fused connection unit leading to a flex outlet unit or to a 13A socket.

Cable size for the circuit

A ring circuit (and its spurs, if any) is wired in 2.5mm² cable, usually 2-core and earth PVC-sheathed. Although this size of cable has a current rating lower than the protective fuse or MCB rating of 30A, each socket is effectively served by two cables, thus splitting the load. Socket outlets and fused connection units on spurs are similarly wired with 2.5mm² cable, but the radial circuit for a fused spur can be wired in 1.5mm² 2-core and earth cable *after* the fused connection unit.

Disposition of socket outlets

Where a house has more than one ring circuit the total number of socket outlets should be distributed as evenly as practicable over the circuits. Usually one circuit supplies the first floor rooms in a two-storey house, a second circuit supplying the ground floor. Wherever possible, the kitchen should also have its own ring circuit, because of the heavy load imposed by kitchen appliances such as washing machines, tumble driers, dishwashers and kettles.

Radial power circuits

A radial power circuit is a circuit supplying a number of 13A socket outlets and usually some fixed appliances from fused connection units (see Fig 93).

The function of a multi-outlet radial power circuit is the same as that of a ring circuit. The only differences are that the circuit cable ends at the last outlet on the circuit and the maximum permitted floor area containing the outlets and therefore supplied by the one radial circuit is less than for a ring circuit. The actual permitted floor area is based on the current rating of the radial circuit.

The 20A radial circuit may supply any number of 13A socket outlets within a floor

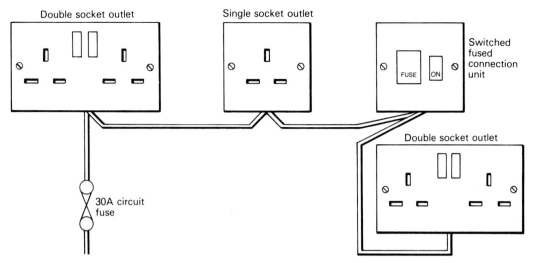

Fig 93 A 30A radial power circuit supplying two double 13A socket outlets, one single and one fused connection unit for a fixed appliance.

area of 20m². The cable of this circuit is 2.5mm² which is the same as that used on a ring circuit. The radial power circuit protection can be an MCB or either a cartridge or rewirable fuse.

The 30A radial circuit may also supply any number of socket outlets and fixed appliances but within a floor area of 50m². The wiring must be in 4.0mm² cable and the circuit excess current protection must be either an MCB or a cartridge circuit fuse, *not* a rewirable fuse.

Other circuits

A circuit supplying a single socket outlet is a radial circuit – an example of this would be a socket outlet for a freezer or a fridge/freezer from the non RCD-protected side of a split-load consumer unit, where all the main socket outlet circuits are RCD protected.

Other radial circuits include circuits for showers, cookers and immersion heaters, all of which are covered in detail elsewhere in this book.

Table 7: **Socket outlet provision**
(Recommendations of the Electrical Installation Industry Liaison Committee)

Room	No of twin sockets
Kitchen	4 [1]
Living room	6
Dining room	3
Double bedroom	4
Single bedroom	3–4 [2]
Landing/stairs	1
Hall	1
Garage	2 [3]
Store/workroom	1
Central heating boiler	1

[1] Not including sockets in cooker control units and not including fused connection units for fixed appliances
[2] Use the higher figure if room is used as study or bed-sitting room
[3] Must be protected by high-sensitivity (30mA) residual current device (RCD)

14: Mainswitches and earthing

Consumer units

A consumer unit is a fuse distribution board and a double-pole main isolating switch contained in the one casing. The unit has a number of ways, termed fuseways, each fuseway supplying a separate circuit. Generally therefore a home electrical installation of six circuits is supplied from a 6-way consumer unit, one of eight circuits from an 8-way consumer unit, a 10-circuit installation from a 10-way consumer unit and so on. Ideally when a house is wired the consumer unit fitted should have at least one spare fuseway to provide for future circuits. Each fuseway, except any spare ways, is fitted with a fuse unit of the appropriate current rating for the circuit.

A fuse unit consists of a base and a detachable or removable fuse carrier containing the fuse element, which is either a fuse wire of a rewirable fuse, or a cartridge of a cartridge-type fuse. Alternatively, instead of fuses, miniature circuit breakers (MCBs) may be fitted (see Fig 94). Some makes of consumer units will accept either fuses or MCBs, others will accept only MCBs.

In addition to the fuse units and the double-pole isolating switch a consumer unit has two multi-terminal blocks; one is the neutral block to which the neutral conductors of all circuits are connected, the other an earth terminal block to which the the individual earth conductors of all circuits are connected, plus the main earthing conductor which is run out of the unit and connected to the consumer's earthing terminal situated near to the electricity company's meter (see Fig 95). Where the electricity company is unable to provide earthing facilities, such as in rural areas where the mains are run overhead, the main earthing conductor from the consumer unit is connected to an earth rod driven into the ground and a residual current device fitted (in the consumer unit or separately between the meter and the consumer unit) to protect all the circuits.

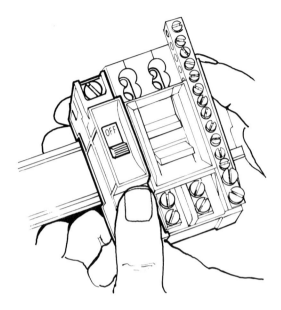

Fig 94 Clipping an MCB on to the top-hat section in the MK 'Sentry' consumer unit. The double-pole isolating switch of the consumer unit is on the right of the MCB.

Isolating switch

The isolating switch in most consumer units is a conventional double-pole mainswitch having a current rating of 60, 80 or 100 amps. The current rating of the switch is largely determined by the size (number of ways) of the consumer unit.

An alternative is to have a 100mA residual current device (RCD) as the isolating switch, providing RCD protection to all circuits. Split-load consumer units have a 30mA RCD protecting some of the circuit ways, with either a double-pole isolating switch or a 100mA RCD as the isolating switch.

Spare fuseways

Any spare fuseways in a consumer unit are fitted with a blanking plate instead of a fuse unit or MCB so that when a circuit is added the correct current rating of fuse unit or MCB can be fitted.

Mainswitches and earthing 179

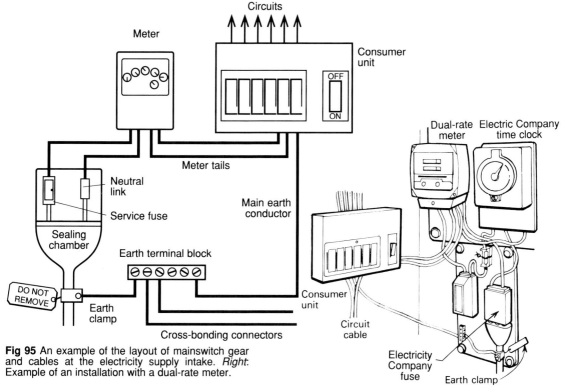

Fig 95 An example of the layout of mainswitch gear and cables at the electricity supply intake. *Right*: Example of an installation with a dual-rate meter.

Photo 107 Fixing the double-pole mainswitch to the top-hat section of a consumer unit (*MK*).

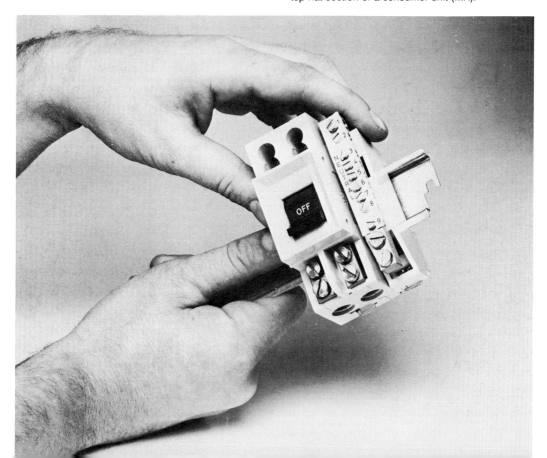

180 Mainswitches and earthing

Choosing a consumer unit

First decide whether you want fuses or MCBs in the consumer unit. MCBs are superior to fuses and should be the first choice. If MCBs, there is much to be said for a consumer unit designed expressly for these. If fuses are chosen the first choice should be the cartridge type as they are superior to rewirable fuses. (See Fuses, page 11).

Next decide on the number of fuseways required, adding two spare ways to allow for future extensions. Choose a consumer unit fitted with an 80A or 100A mainswitch rather than a 60A. Decide whether the isolating switch is to be a conventional mainswitch or an RCD, remembering that the latter can be installed as an additional item supplying selected circuits (see *Residual current devices*, page 187).

Installing a consumer unit

Whether the installation is an entirely new one or, which is more likely, an existing installation being up-dated or rewired, the procedure for fitting a new consumer unit is basically the same.

If an existing installation the new unit will either be replacing another consumer unit which is too small for the expanding installation or it will be replacing a miscellany of old-fashioned mainswitches and fuse units of various types.

Materials and equipment

The basic requirements are: the consumer unit itself containing fuseways, terminals and the double-pole mainswitch of the appropriate current rating; one fuse unit or MCB for each circuit and of the appropriate current ratings which include 5A, 15A, 20A, 30A and where required 45A, it being essential that the model of consumer unit chosen will accept a 45A fuse unit, where fuses and not MCBs are chosen (MCBs have different ratings from fuses – typically 6A, 10A, 16A, 20A, 32A, 40A and 45A); one red and one black PVC-insulated and -sheathed single-core 16mm^2 cable, each about 1m in length, to run from the mainswitch terminals to the elec-

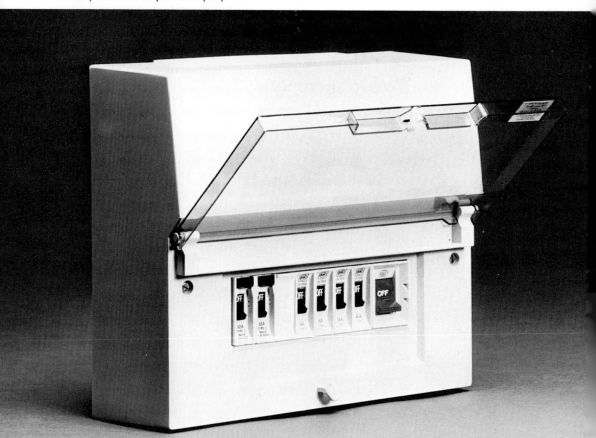

Photo 108 6-way consumer unit with miniature circuit breakers. The two circuits on the left (for socket outlets) are both RCD protected (*MK*).

Mainswitches and earthing 181

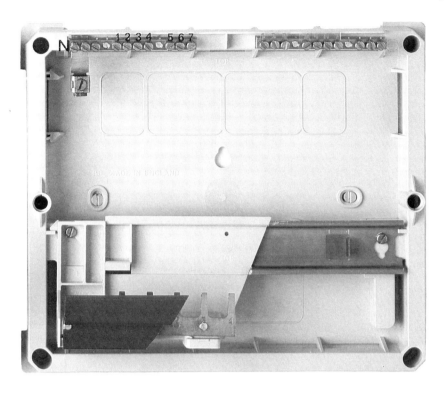

Photo 109a Schematic of consumer unit, showing busbar and DIN rail options. The neutral and earth terminal blocks are at the top (*Ashley & Rock*).

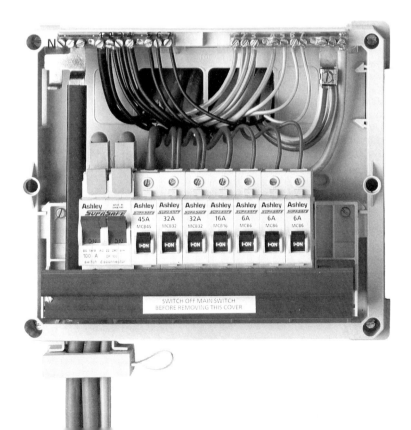

Photo 109b The 7-way consumer unit wired up with MCBS (*Ashley & Rock*).

tricity company's meter – these are termed meter tails; one length of 6.0mm² single-core green/yellow PVC-insulated earthing cable of sufficient length to run from the earth terminal block in the consumer unit to the earthing terminal close to the meter or wherever the terminal is situated; green/yellow PVC sleeving for the earth conductors of the circuit cables within the consumer unit; screws and wall plugs.

Fixing a consumer unit

If the new consumer unit replaces an existing one or an arrangement of mainswitches and fuses it is first necessary to remove this old equipment. The first job is to ask the electricity company, in writing, to withdraw the service fuse and disconnect the meter leads from their meter and to leave the fuse withdrawn until notice is given to replace it. If the change-over is fairly simple the service fuse can be removed in the morning and replaced later the same day.

Removing an old consumer unit

With the electricity supply having been disconnected by the electricity company, remove the cover of the old consumer unit and disconnect the two meter tails and the main earth conductor.

Remove each fuse unit in turn and disconnect the circuit wires, making a note of the circuit and attaching a label on the cable. Remove the cables from the unit and remove the unit from the wall leaving only the circuit cables.

Where the existing arrangement comprises a number of switch and fuse mains units, disconnect, label and remove each in turn. Before these can be replaced by a single unit it is advisable to check that the circuit cables will all reach the terminals of the new consumer unit. Adjusting the position of the new unit can affect this, with the cables now reaching the new consumer unit.

Make good the holes in the wall and any damage to the plaster. Repaint or repaper the wall if required.

Installing the new consumer unit

The method of preparation and fixing the new consumer unit depends upon the make and model.

Most consumer units are available either with an all-insulated plastic mounting box or with a metal-clad mounting box, both designed for surface mounting to a wall or a purpose-made plywood mounting board.

Both metal and plastic mounting boxes have a choice of knock-outs in the sides and the back for taking the circuit cables, the meter tails and the main earthing conductor. Where a metal mounting box is used, grommets should be fitted to the knock-out holes to protect the cables from chafing on the metal.

Hold the consumer unit against the wall and mark the positions of the fixing holes. Drill and plug the holes for No 10 or No 12 wood screws.

Before fixing the unit it is better to connect the meter tails and the main earth conductor. Remove about 30mm of outer sheathing from the ends of the red and the black cable and about 25mm of insulation. Thread the ends of the leads into the casing. Connect the red lead to the MAINS 'L' terminal and the black to the MAINS 'N' terminal. Connect the end of the 6.0mm² earth wire to the end terminal of the earth terminal block. With all fuse units or MCBs removed (if fitted on purchase), fix the unit to the wall using the wood screws.

Connecting circuit cables

The red (live) wires of the circuit cables are connected to the respective fuseway terminals or to the terminals of the MCBs, the black wires to the neutral terminal block and the sleeved earth conductors to the earth terminal block.

The live wire (or two wires, if a ring circuit) of the circuit having the highest current rating is connected to the fuseway or MCB nearest the isolating switch, and those of 5A current rating to the fuseways farthest from the mainswitch. Therefore a 45A fuse unit or MCB will be fixed next to the main isolating switch. If no 45A circuit is being fitted at the moment, it would be a good idea to leave this 'way' blank to allow for a 45A shower circuit to be fitted in the future.

Connect one circuit in turn, remembering that each ring circuit has two cables which have two red, two black and two earth conductors, with two wires of the same

Mainswitches and earthing 183

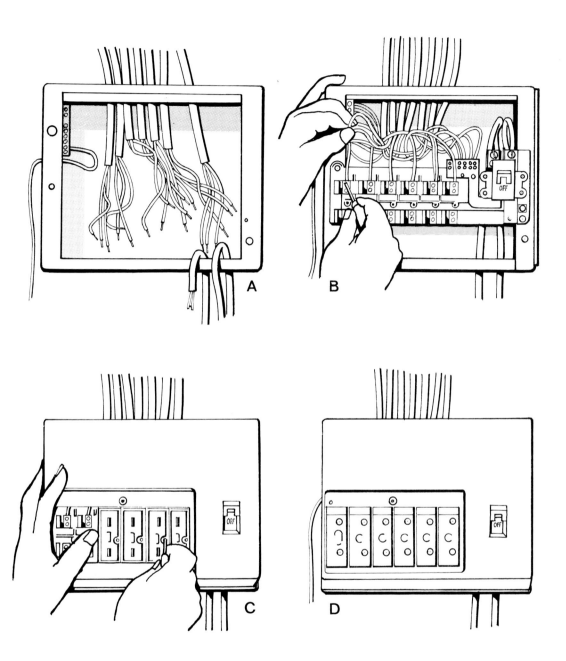

Fig 96 Installing a 6-way consumer unit: **A** Secure the unit to the wall and pass the cables in through knockouts if necessary; **B** Connect up the circuit wires in order; **C** Secure the fuse carriers or MCB holders; **D** Fit the fuses or MCBs and then put on the safety cover.

colour going to each respective terminal. Where there are two (or more) ring circuits make sure that it is the two ends of the same cable which are connected to each terminal, not one end of each of two ring cables. As each circuit is connected jot down its details in the correct order.

The black wires and the earth wires of each circuit should be connected to the individual terminals of the neutral and earth terminal banks in the same order as the cables in the fuseways or MCBs so that the complete circuit can readily be disconnected at any time if necessary for testing.

Where the consumer unit is fitted with MCBs instead of fuses the red live circuit wire (or wires) of each circuit is connected to the terminal of the MCB and the MCB is then fitted to the consumer unit. An exception is with a consumer unit which accepts fuses and MCBs. This type of consumer unit has the conventional fuseway terminal with the MCBs being inserted later in the same way as fuse units.

Assembling the consumer unit

With all circuit cables now connected the fuse units are fixed in position and the covers replaced. At the same time circuit details are entered on a label pasted inside the covers or on labels underneath the fuses/MCBs. Where the manufacturer offers other accessories for consumer units (bell transformers, time switches and time delay switches), these should be fitted at the same time.

With the consumer unit now installed and all cables connected, the electricity company is called in to replace the service fuse.

Installing a switchfuse unit

A switchfuse unit is, in effect, a one-way consumer unit. The procedure for fixing and connecting it is the same as for a multi-way consumer unit except there is only one circuit and therefore only one fuseway or MCB.

The two meter tails are connected to the mains terminals. Although the unit supplies only one circuit which could be of any current rating from 5A to 45A, the meter tails have to be of the same size as for a consumer unit – which normally is 16.0mm². The reason for this size of meter tails is that they must not only be adequate to carry the circuit current but they must be of the appropriate

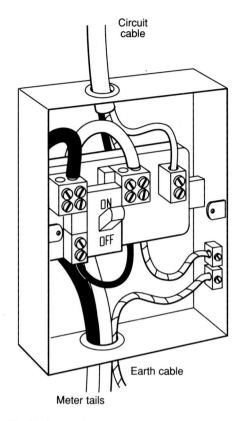

Fig 97 Connection of the meter tails and circuit cable in a switchfuse unit.

size for the protecting fuse which in this case is the electricity company's service fuse (up to 100A).

A combined RCD/MCB unit (sometimes known as an RCBO) can be used in a switchfuse unit enclosure and is a good choice where RCD protection is needed for a single circuit – such as a garden circuit.

Connection of meter tails

Normally an electricity company will not connect more than one pair of meter tails to its meter, which for one thing may not be able to accommodate more than one pair in the terminals.

Where, therefore, a switchfuse unit is added to an installation when a shower unit, garage or an outdoor circuit is installed, special arrangements have to be made for its connection to the mains.

The usual arrangement is to use a service connector box (also known as a 'splitter unit' or a 'Henley box'). This is a moulded plastic box with a detachable cover and which contains two 5-terminal blocks, one for the live tails and one for the neutral tails. The box is fixed to the wall close to the meter and the existing consumer unit. At the moment, there will be a pair of tails (live and neutral) connecting the consumer unit to the meter. With a service connector box, one pair of tails connects from the meter to the service connector box and two further pairs from the connector box to the existing consumer unit and the new switchfuse unit (or second consumer unit). Up to three additional switchfuse units (or consumer units) can be fitted if required.

You will have to provide the two extra pairs of tails (available in 1m lengths) and then ask the electricity company to come in and fit them for you, once the switchfuse unit is in place and wired to its circuit. A new earth conductor will need to be taken from the main earthing point into the new switchfuse unit.

Earthing and bonding

Every circuit of a domestic wiring installation must be earthed – that is, must have an earth conductor running throughout the circuit. With PVC-sheathed cable, as used in house wiring, the earth conductor is a bare wire enclosed in the sheathing between the two (or with 2-way lighting cable, three) PVC-insulated current carrying conductors. Wherever this bare wire is exposed inside mounting boxes for accessories, it should be covered with green/yellow sleeving.

The earth conductors are connected to the earthing terminal block in the consumer unit and run to every switch, ceiling rose, socket outlet and other wiring accessory in the house. The earthing terminal block in turn is connected to the consumer's main earthing point, normally provided by the electricity company. Where no earthing point is provided, it may be necessary to fit an RCD and to run an earth conductor from an earth rod (electrode) driven into the ground (see Fig 99).

It is not always understood by consumers that it is they and not the electricity company who are responsible for earthing, though a company usually provides the facilities. An earth rod is copper-coated steel, needs to be

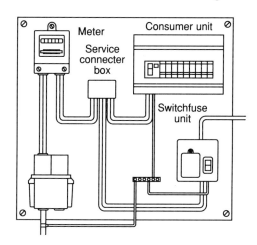

Fig 98 A service connector box fitted to supply a switchfuse unit.

at least 1.2m long, and has an earth terminal on the top for connecting the earthing lead. A label is affixed to the terminal which reads 'SAFETY ELECTRICAL CONNECTION – DO NOT REMOVE'. The top of the rod should be in an enclosure to protect the connection and to provide access.

Purpose of earthing

The purpose of earthing in an electrical installation is to blow the fuse or to trip a miniature circuit breaker (MCB) when the live side of the electric wiring or that of an appliance comes into contact with exposed metalwork such as the casing of an appliance, the body of a kettle, toaster or electric iron.

If in these circumstances the metal casing of the appliance was not earthed, a person touching the live metal frame and at the same time in contact with a water pipe or other earthed metalwork, or even standing on a concrete floor, would receive a severe electric shock which could be fatal. This is because the neutral side of the supply is earthed and in effect a person in contact with the live metalwork and earthed metalwork is touching both sides of a 240V supply.

When the casing of the appliance is earthed, as it must be in the interests of safety, and a live wire comes into contact with the metalwork there is an immediate surge of very heavy current and the fuse blows or the MCB trips, as it does when the

186 Mainswitches and earthing

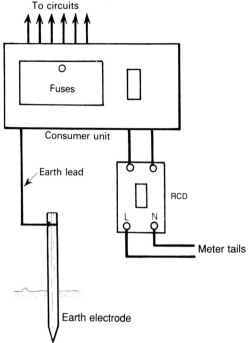
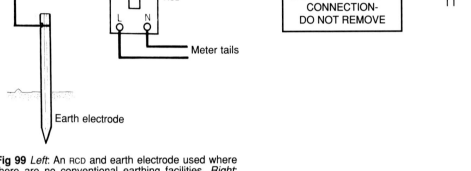

Fig 99 *Left*: An RCD and earth electrode used where there are no conventional earthing facilities. *Right*: The earth electrode must be in an accessible enclosure and fitted with a warning label.

live and neutral wires come into contact and cause a short circuit. With the fuse blowing or the MCB tripping and the electricity cut off, the metalwork is safe to touch. The fuse having the lowest current rating blows – which for an appliance is usually a plug fuse, and for a circuit is the circuit fuse or MCB.

Bonding

In addition to providing earthing throughout the electrical installation it is also necessary to connect the gas and water mains pipes to the electrical earthing. This is known as equipotential bonding. It may also be necessary to bond metal items in the bathroom and elsewhere to earth. This supplementary bonding is particularly important with protective multiple earthing – see next page.

Equipotential bonding

Fit an earthing clamp to a cleaned portion of the mains water pipe as close as possible to the point where it enters the house and on the street side of the indoor stop cock. Fit an identical clamp to the mains gas pipe on the house side of the gas meter (see Fig 100).

From one clamp to the other run a length of 6mm^2 single-core green/yellow PVC-insulated cable following the shortest practicable route. Connect one bared end of the cable to one clamp and the other end of the cable to the other clamp. From the clamp on the pipe nearest to the electricity mains supply, run a length of the same cable to the consumer's earth terminal and connect it to that terminal.

Supplementary bonding

Bathrooms pose a particular hazard, because the body is in contact with water, and it is necessary to bond all extraneous metalwork – including radiators, metal baths and pipework – with the earths of electrical equipment, such as showers or heaters.

Where an electric shower has been fitted in a bathroom a 4mm^2 single-core earthing cable can be connected to the earth terminal of that and taken to earthing clamps fitted to water pipes and central heating pipes, to metal waste pipes, to towel radiators, to the

Mainswitches and earthing

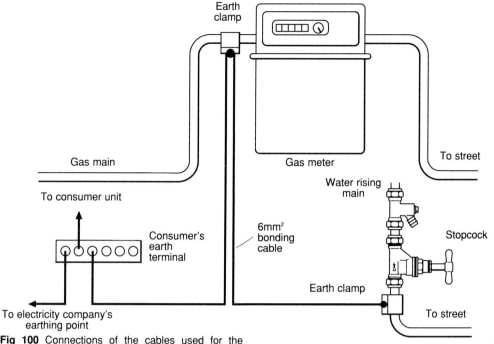

Fig 100 Connections of the cables used for the necessary bonding to earth of the mains water and gas services.

metal cases of electrical appliances and to the earthing terminal fitted on to the bath. Otherwise, a 4mm² single-core earthing conductor should be taken to the bathroom all the way from the main earthing point. Metal sinks in kitchens should also be earthed. If plastic connectors are used to join copper pipe, an earth wire should be fitted either side of the connector to provide earth continuity. Special kits are available for doing this.

Where Protective Multiple Earthing (PME) has been used, the electricity company will insist that you bond all service pipes and extraneous metalwork in bathrooms (and elsewhere) and you should check with them on the size of conductor required – normally this will be 6mm². With PME, the earthing terminal is actually connected to the neutral side of the mains supply, but the earth and neutral conductors in the circuits within the house remain separate.

Residual current devices

A residual current device (RCD) is a double-pole switch which trips (switches off) automatically when electric current leaks to earth in a sufficient amount to operate the tripping mechanism. The amount of earth leakage current required is always substantially less than is required to blow either a circuit or a plug fuse. It works on the principle that when there is no current leakage in the installation the amount of current flowing out through the neutral wires is identical to that flowing in through the live wires and therefore a current balance in the RCD is maintained.

When a fault occurs and current leaks into the earthing system the amount of current flowing out through the neutral is less than in the live wire, the difference being that of the earth leakage – a current which could be enough to cause a fire or to give anyone in contact with earth a fatal electric shock.

This out-of-balance situation is detected by the residual current device causing the tripping coil to be energised which operates the trip. The amount of current required to trip an RCD varies with the model. Most high-sensitivity RCDs have a tripping current of just 30mA (10mA versions are also available).

In general 30mA sustained for a few milliseconds (which is the time taken for the RCD to respond) is less than required to elec-

188 Mainswitches and earthing

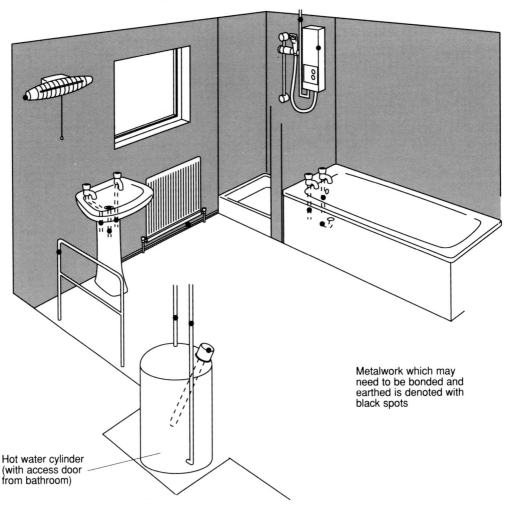

Metalwork which may need to be bonded and earthed is denoted with black spots

Hot water cylinder (with access door from bathroom)

Fig 101 Bonding to earth of extraneous metalwork.

trocute the normal healthy person. Installing a high-sensitivity RCD therefore not only gives greater protection from fire and shock because the RCD operates more quickly than a fuse and needs less current to do so, but it also affords greater protection to a person accidentally touching a 240V live wire or contact.

Where an RCD is fitted in or before the consumer unit to protect the whole installation, this will normally be 100mA to avoid the problem of 'nuisance' tripping.

Installing an RCD

An RCD can be installed to protect the whole house, to protect particular circuits, to protect just one circuit, to protect a socket outlet or to protect appliances plugged into a socket outlet.

Whole-house RCD

As illustrated on page 186, an RCD can be fitted between the meter and the consumer unit to protect the whole house (in Fig 99 it is being used in conjunction with an earth electrode).

As explained on page 178, the main isolating switch in the consumer unit can be replaced with a (100mA) RCD, which will also protect all the circuits in the house.

Circuit protection

An RCD can be fitted into the cable of any one

Mainswitches and earthing 189

Photo 110 Residual current devices, suitable for use in a consumer unit or a separate enclosure (*Crabtree*).

circuit to protect everything that is on that circuit (with a ring main, it must be connected into *both* cables). But if you want more than one circuit protected, it would be tedious and expensive to fit an RCD to each one, so the answer here is a split-load consumer unit with a 30mA RCD protecting socket outlet, shower and cooker circuits, for example. The lighting circuits do not have RCD protection (the likelihood of fire and electric shock is much lower here), which means the lights will stay on if the RCD trips. Where the isolating switch in a split-load consumer unit is a second RCD, this will have

Photo 111 RCD-protected metal-clad socket outlet with two sockets (*PowerBreaker*).

a 100mA rating and usually be time-delayed so that the 30mA RCD always trips first.

Socket outlet protection

It is possible (but expensive) to fit an RCD in the cable(s) leading to a socket outlet. But a better solution is either to protect the whole circuit or to replace the (double) socket with an RCD-protected socket – usually giving only a single socket outlet. All sockets which can be used for powering equipment outside should be RCD protected.

Appliance protection

An RCD socket will protect all the equipment which is plugged into it, but not any other sockets on the circuit. An RCD adaptor plugs into a socket and protects whatever appliance is plugged into it, so can be moved around from socket to socket. On the other hand, it could get damaged and you have to remember to use it each time. An alternative for vulnerable equipment, such as lawnmowers or hedgetrimmers, is to fit the appliance with an RCD-protected plug so that it is always protected whichever socket it is plugged into.

190 Mainswitches and earthing

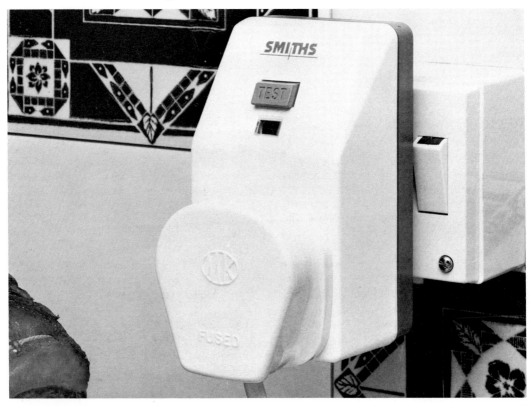

Photo 112 RCD adaptor fits between the appliance plug and the socket outlet (*Smiths Industries*).

Photo 112a RCD plug for attaching to equipment used outdoors (*PowerBreaker*).

Fitting RCDs

An RCD-protected plug is fitted in the same way as a normal plug, whilst an RCD-protected adaptor needs only to be plugged into a socket outlet. An RCD-protected socket is wired up in exactly the same way as a normal socket outlet, whilst an RCD in a split-load consumer unit is connected within the consumer unit itself.

The only problem comes with wiring a circuit RCD connected into the cable (or cables) of the circuit. The live and neutral connections are straightforward enough – one each to one pair of terminals and one each to another – but inside the enclosure (which is normally bought separately), you will need to fit a terminal connector block to take the two earth wires as there is no terminal for them on the RCD itself.

All RCDs come with a test button which should be tried regularly – every time with a socket or adaptor RCD.

15: Electricity supply requirements

Obtaining an electricity supply

When moving house it is necessary to contact the local office of the electricity company which serves your new area in order to obtain a supply of electricity from the date of removal and to arrange the tariff. This procedure is necessary whether a new house or a change of occupancy.

A form of application needs to be completed on which you have to state that the maximum power will not exceed a certain amount (typically 14kW/25kW). You also have to choose which of two tariffs you want – general purpose tariff or Economy 7 tariff. Information concerning any new wiring work must also be submitted. The electricity company representative reads the meter to ensure that you are not charged for any electricity consumed before the date of your occupancy. If the house has been unoccupied for some time before you move in, the electricity company may want to carry out some tests of the wiring.

Electricity tariffs

The tariff governs how much you pay for your electricity. There are two principal tariffs available to domestic consumers – the general purpose tariff and the Economy 7 tariff. Which one you should choose depends on how (or, more to the point, when) you use your electricity.

The general purpose tariff has two components: a quarterly standing charge, (which covers the cost of providing and maintaining metering and cabling and of sending out electricity bills) and a unit charge of so many pence per unit (kWh). All units of electricity used are charged for at the same rate; the standing charge depends on the type of meter you have.

The Economy 7 tariff has the same two components, but the quarterly charge is slightly higher and there are two unit charge rates – one for electricity used during the day and one for electricity used at night. The daytime rate is slightly higher than the cost

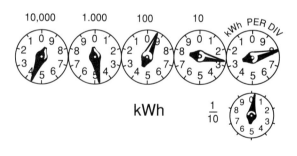

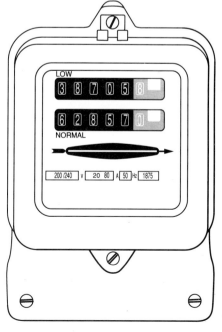

Fig 102 *Above*: To read a dial meter, remember that adjacent dials rotate in opposite directions. Ignore the tenths dial (only there for testing the meter) and, starting with the 10,000 scale, write down the number the pointer has passed. In this case the correct reading is 44928 – note that the reading on the 1,000 scale is *not* 5 as the pointer on the 100 scale has not yet reached 0. *Right*: A two-part digital meter can be read directly, the top row for night-time use, the bottom for daytime use.

192 Electricity supply requirements

under the general purpose tariff, but the night-time rate (available for seven hours, typically between 00.30 and 07.30) is less than half the normal daytime rate. The timing of the Economy 7 period varies, but it is based on Greenwich Mean Time, so will be an hour later in the summer.

The Economy 7 tariff is designed primarily for users of storage heaters and electric immersion heaters, but if you use electricity at any time of the year to heat your water or can arrange for heavy users of electricity (like washing machines, tumble driers and dishwashers) to run at night, you will be better off under this tariff. Note that *all* electricity used at night is at this cheap rate, including outside lights and fridges and freezers.

Converting to Economy 7 usually means only changing the meter for one with two dials plus a timeclock (a service usually offered free by the local electricity company); if you are installing storage heaters for the whole house, you will also need a second consumer unit, connected to the meter (by the electricity company) in such a way that it operates only during the cheap night-time period.

Saving money on electricity

Using electricity sensibly is the best way of saving money. With lighting, for example, you can save by switching lights off when not required, but a better way is to make use of fluorescent lighting and low-voltage lights. A 20W fluorescent tube or a 15W compact fluorescent lamp will give the same output as a 100W light bulb, but for a fraction of the cost. You have to pay more for the lamps (and, if necessary, the fitting) but the fact that fluorescent lights last up to seven times longer than normal light bulbs makes them much more economical. Low-voltage lighting (see page 79) shows a similar cost benefit. Using dimmer switches is not a good way of saving money on lighting.

Immersion heaters are among the biggest wasters of electricity, mainly from heat loss. Using a pre-insulated cylinder (or fitting a 100mm insulating jacket to an existing cylinder) will cut down consumption considerably, but the best way of saving money is to use an immersion heater timer or an Economy 7 controller to bring on the immersion heaters only when you want them or at night during the Economy 7 period (assuming you have converted to Economy 7).

Modern storage heaters have both input and output controls, so you can adjust the amount of heat being taken in and the rate at which it is given out, and some have the ability to carry heat not wanted one day into the next day. Using full-price electricity during the day for heating can be expensive, but most modern heaters are fitted with thermostats which will reduce electricity bills and some have timers so that they are not left on wastefully.

Electricity consumption

To ascertain an appliance's rate of consumption of electricity, you calculate on the basis that a 1kW heater burns one unit (kWh) of electricity each hour it is in use. The rate of consumption of appliances having a loading less than 1kW is less than one unit per hour; and those having loadings more than 1kW burn more than one unit per hour.

While it is useful to know how much electricity is consumed by an appliance during any period, it is probably more useful to know what an appliance will provide for each unit of electricity consumed (see Fig 103). The table below provides this information for many of the more popular appliances used in the home:

1 Fridge-freezer: one day's use
2 Iron: over 2 hours' ironing
3 Blender: 500 pints of soup
4 Toaster: 70 slices of toast
5 Kettle: 12 pints of water
6 Electric shower (8kW): $7\frac{1}{2}$ minutes' showering
7 Television set: 6-9 hours' viewing
8 Waste disposal unit: 25kg (50lb) of kitchen waste
9 Can opener: several thousand cans
10 Carving knife: 200+ joints sliced
11 Cooker hood: 10 hours' operation
12 Washing machine: weekly wash for a family of four uses 4–5 units
13 Cooker: one week's meals for family of four uses 17 units
14 Extractor fan: 24 hours' operation
15 Vacuum cleaner: 2 hours' cleaning
16 Hot tray $1\frac{1}{2}$ hours' use

Electricity supply requirements 193

Fig 103 A range of electrical appliances used in the home.

17 Blanket: overblanket, 2 full nights; underblanket, 7 nights
18 Hair rollers: over 20 hair-do's
19 Hair dryer: 12 ten-minute drying sessions
20 Fan convector heater (2kW): 30 minutes' heating
21 Tea maker: 35 cups of tea
22 Microwave oven: 2 × 1.4kg (3lb) joints of beef.

Connection of new wiring

When an electrical contractor rewires a house or carries out significant new wiring work, he completes an installation certificate which states that the electrical installation has been inspected and tested and complies with the IEE wiring regulations, any departures from the regulations being noted.

When the DIY householder carries out new wiring work or a major alteration such as installing a ring circuit, installing a cooker circuit or a shower unit, or running electricity outside to a detached garage, shed or greenhouse, he cannot make this certification unless an electrical contractor has been employed to do the inspection and testing necessary, since specialised testing equipment is needed.

However, the form can still be completed, asking the electricity company to make the necessary inspection and tests, for which a fee is payable.

Wiring inspection and tests

The IEE wiring regulations (see opposite) specify the inspection and tests which should be carried out on new, existing or modified wiring installations.

The inspection consists of looking for things like the correct choice and installation of equipment, the correct connection and identification of conductors, the correct sizing of conductors (including earth bonding conductors), the correct positioning of single-pole switches (in the live not the neutral side of the circuit), the correct connection of socket outlets and lampholders, correct earthing, bonding and insulation, the presence of isolating switches and the accessibility of equipment and wiring accessories.

The tests which the electricity company or a professional contractor will carry out include:
(i) earth continuity – checking that all earth conductors and bonding conductors are effective;
(ii) ring circuit continuity – checking that all three conductors effectively complete the ring circuit;
(iii) insulation resistance – applying a DC voltage of 500V to check for the insulation resistance between conductors (carried out with the circuit disconnected from the supply and with any vulnerable electronic equipment disconnected);
(iv) polarity – to check the wires (eg to socket outlets) are connected the correct way round and that single-pole switches are only in the live side of the circuit;
(v) earth fault loop impedance – to check the effectiveness of the main earthing (where an earth electrode is fitted, this has to be tested, too);
(vi) RCD operation – to verify that RCDs operate properly.

DIY inspection and test

A DIY electrician can carry out most of the visual inspection checks above, though the electricity company will want to check things like the size of earth bonding conductors.

When it comes to testing, the two main tests which the amateur can do are continuity and polarity of a new or existing circuit. A new circuit should be tested *before* it is wired up to the mains: continuity of a ring circuit, for example, can be tested at the ends of the two cables, checking with a continuity tester or multimeter that the live, neutral and earth wires form a complete circuit and do not cross over; polarity means using a 'flying lead' to check the connections at each socket outlet. Flying leads should also be used to test other circuits.

To test an existing circuit, a socket outlet tester can be used with the electricity turned on. This will indicate the following faults: missing earth, live/neutral reversed, live/earth reversed, missing or faulty neutral, missing or faulty live.

Important wiring regulations

Regulations which affect the home electrical installation are the Regulations for Electrical

Photo 113 This plug-in tester instantly checks whether a 13A socket outlet is correctly connected to the circuit wires (*Maestrolite*).

Installations (1991) published by the Institution of Electrical Engineers and usually referred to by the short title 'IEE wiring regulations'.

The regulations are continually being amended and occasionally completely revised. The first edition appeared in 1882 and the current edition (16th) was published in 1991. Although not a statutory document, the regulations are recognised as a code of practice by all bodies associated with electrical installations as well as government departments and safety organisations. In England and Wales, compliance with the IEE wiring regulations is not enforceable by law unless written into a contract but non-compliance with those regulations affecting safety will contravene the statutory electricity supply regulations which could mean an electricity company refusing to connect up an installation. In Scotland, the wiring regulations are part of the building regulations, so are a mandatory requirement.

Electricity companies have no power to refuse connection where the wiring has been carried out by a DIY electrician, providing it is safe, but from the safety aspect and every other aspect DIY wiring (and, for that matter, wiring by a contractor) should comply with the regulations. A principal requirement of the regulations is that good workmanship and proper materials should be used throughout.

Choosing a contractor

Electrical contractors whose names are on the register of the NICEIC (National Inspection Council for Electrical Installation Contracting) and those who are members of the Electrical Contractors Association are required as a condition of continuing registration or membership to comply with the regulations for all wiring work they carry out. Therefore when employing a contractor, ensure that he is on the NICEIC register as are the contracting departments of all electricity companies. Work carried out by members of the Electrical Contractors Association is guaranteed.

16: Safety in home electrics

When working on wiring and appliances of the home electrical installation, the element of safety is of primary importance. The electricity of the supply at 240V AC is lethal. Bad wiring and poor workmanship can lead to fire and shock as also can mishandling and the misuse of electrical appliances. The points on safety listed below should be taken notice of and observed at all times.

General

(i) Do not attempt any electrical work, no matter how small, if it is beyond your capabilities.
(ii) Turn off the main isolating switch and remove circuit fuses (or switch off MCBs) before working on the fixed wiring.
(iii) Pull the plug out of its socket before working on a portable appliance or a portable lamp.
(iv) Use only high quality cable, wiring accessories and other materials for the work you undertake.
(v) When testing wiring or attempting to locate a fault which requires the electricity to be switched on and off as the checking proceeds, take especial care that the electricity is off when handling live wires and live parts of a circuit or apparatus.
(vi) Ensure that all work will comply with the IEE wiring regulations by carefully following the instructions given in this manual and remembering that 'good workmanship' is a requirement for compliance with the regulations.

Flexible cords

Flexible cords, although a convenient means of connecting portable appliances and lights to the fixed wiring, are the weak spots in an installation. The following points should be carefully noted.

(i) Do not patch frayed flex with insulation tape. Replace the flex at the first sign of wear or damage.

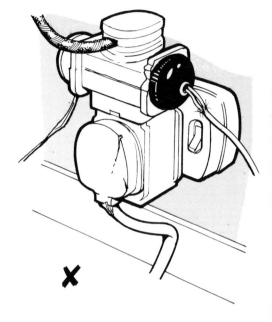

Fig 104 Never use plug adaptors to build up a dangerous 'Christmas tree' effect, possibly resulting in a fire.

(ii) Do not run flex under carpets, rugs or other floor coverings.
(iii) Do not use old twisted twin flex.
(iv) Do not connect 3-core flex to a two-pin plug.
(v) Do not fit 2-core flex to appliances other than double-insulated appliances marked with a double hollow square.
(vi) Do not have unnecessarily long flexes. Keep as short as possible to prevent damage to the flex or accidents to people through tripping over them. Avoid using multi-plug adaptors (see Fig 104) for apart from the risk of overloading a socket outlet such practice usually means long flexes. To avoid the temptation, put in extra sockets..
(vii) Too short a flexible cord on an iron or a kettle can however place undue strain on the flex and prevent the appliance being used safely.
(viii) Ensure that a flexible cord does not drape over a switched-on boiling ring or a gas ring or over a radiant electric heater.

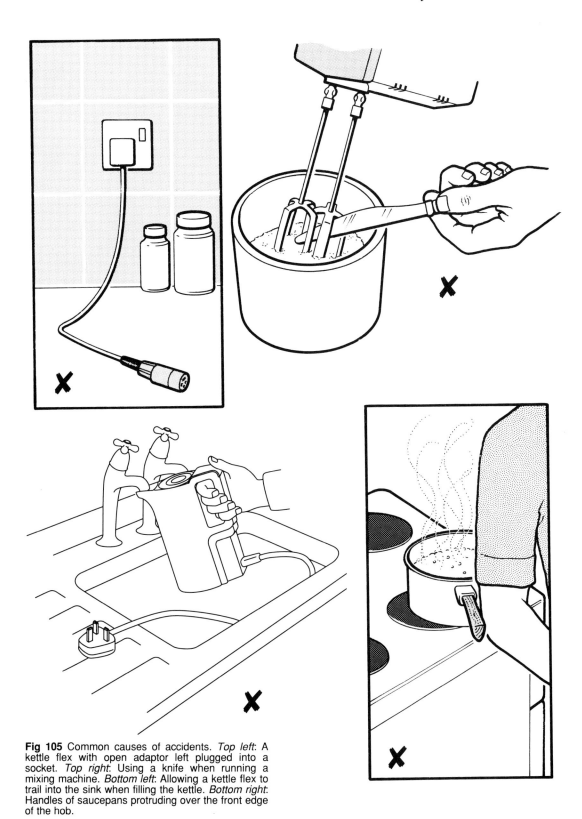

Fig 105 Common causes of accidents. *Top left*: A kettle flex with open adaptor left plugged into a socket. *Top right*: Using a knife when running a mixing machine. *Bottom left*: Allowing a kettle flex to trail into the sink when filling the kettle. *Bottom right*: Handles of saucepans protruding over the front edge of the hob.

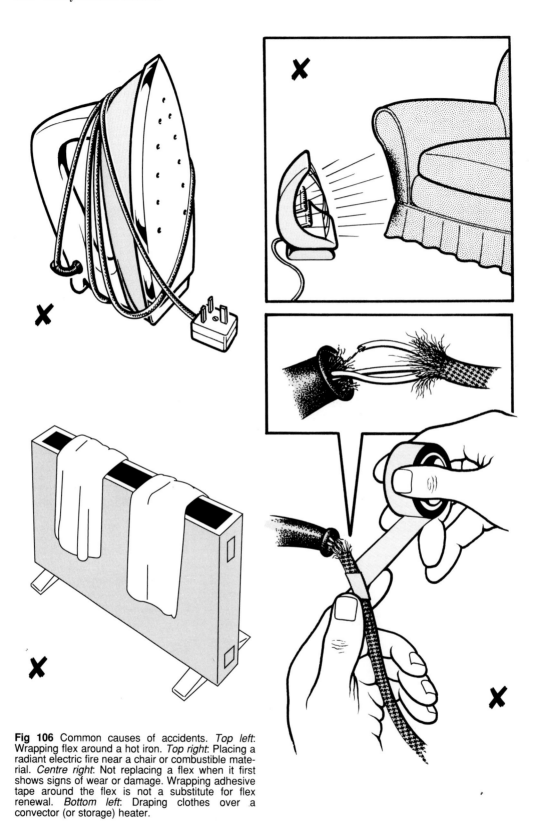

Fig 106 Common causes of accidents. *Top left*: Wrapping flex around a hot iron. *Top right*: Placing a radiant electric fire near a chair or combustible material. *Centre right*: Not replacing a flex when it first shows signs of wear or damage. Wrapping adhesive tape around the flex is not a substitute for flex renewal. *Bottom left*: Draping clothes over a convector (or storage) heater.

(ix) Do not wrap the flex around an electric iron after use until the iron is cool.
(x) Do not connect two or more flexible cords to a ceiling rose designed for only one flex.
(xi) Do not join flex except with a purpose-made flex connector or line cord switch.
(xii) Do not wire more than one flex to a single plug.

Safety in the bathroom

Special precautions are necessary to ensure safety in the bathroom – that is, a room containing a fixed bath and/or shower.

(i) The light fittings should be totally enclosed or where an open lampholder is fitted this must have a deep skirt to shield the metal lampcap.
(ii) No wall switch must be in a position where it can be reached by a person using the bath or shower. It is necessary to use cord-operated ceiling-mounted switches.
(iii) No portable mains-operated appliance may be used in a bathroom nor must any provision be made for using an appliance, which means there must be no socket outlets. The only normal exception to this rule is a shaver supply unit fitted with an isolating transformer.
(iv) All extraneous metalwork must be bonded to earth, see page 186.
(v) In other rooms (eg bedrooms) containing a shower enclosure, wall switches must be at least 0.6m and sockets at least 2.5m from the shower enclosure.

Appliances

Electrical appliances if misused or neglected are sources of danger. Take note of the following points.

(i) Do not attempt any repairs or adjustments to an electrical appliance unless you are conversant with such repairs or have a copy of the maker's repairs manual and are using the correct components.
(ii) Do not stand a radiant electric fire against furniture or fabrics or face it against a wall or woodwork.
(iii) Do not place clothes or materials to dry over a convector heater, over a fan heater or over the outlet grille of a storage heater.
(iv) Always buy plugs that conform to BS1363 and fit the correct rating of fuse. Wire plugs correctly and throw away damaged moulded-on plugs which become faulty.
(v) Do not fix an airing cupboard heater at low level without fitting a clothes guard above the heater.
(vi) Do not touch the elements of a portable electric fire without first pulling out the plug or, if of the fixed type not supplied from a socket outlet, without first turning off the main switch.
(vii) Do not dismantle a cooker for cleaning without first turning off the control switch. Do not line any part of the cooker with aluminium kitchen foil.
(viii) Do not attempt to remove pieces of toast from a toaster using a knife or other metal object. Always pull out the plug *before* cleaning a toaster or removing crumbs.
(ix) Do not use appliances out of doors or in the greenhouse unless they are designed for the purpose and protected by an RCD.
(x) Do not use appliances that are not earthed, with the exception of those which are double insulated and bear the double hollow square sign.
(xi) Do not run an appliance from a lampholder socket using a lampholder adaptor, or from a 2-pin socket apart from a shaver plugged into a special shaver supply unit or shaver socket.
(xii) Buy only BEAB-approved appliances.
(xiii) Always switch off light fittings before changing lamps and do not exceed maximum recommended wattage for bulbs.
(xiv) Never handle plugs or switches with wet hands.
(xv) Unplug kettle before refilling.
(xvi) Turn off television set after use and remove mains plug from socket.
(xvii) Unplug steam iron before refilling with distilled or deionised water.
(xviii) Follow instructions for using electric blankets and have the blankets serviced regularly.

Wiring

(i) Always use the correct sizes and types of cable.
(ii) Do not install a switch, light fitting, socket outlet or other wiring accessory unless it is mounted on the necessary mounting box or backplate or it incorporates a backplate.

200 Safety in home electrics

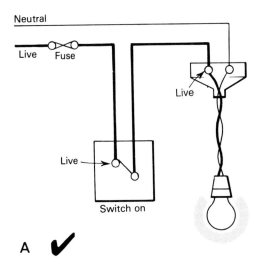

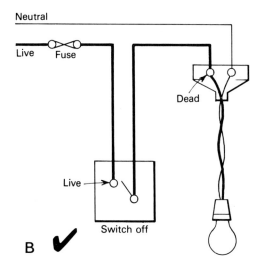

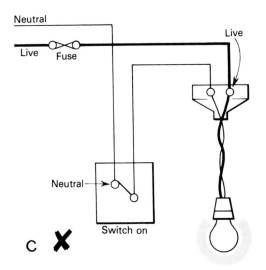

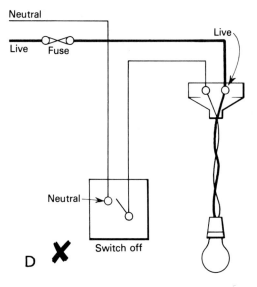

Fig 107 *Top*: A single-pole switch must always be inserted into the live side of the circuit. **A** This is the correct method as the lampholder terminal in **B** is 'dead' when the switch is OFF; *Below*: **C** The incorrect method with the switch in the neutral side of the circuit. When the switch is OFF, the lampholder terminal **D** in the ceiling rose remains 'live'.

Safety in home electrics **201**

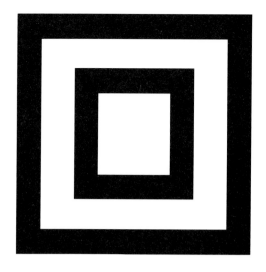

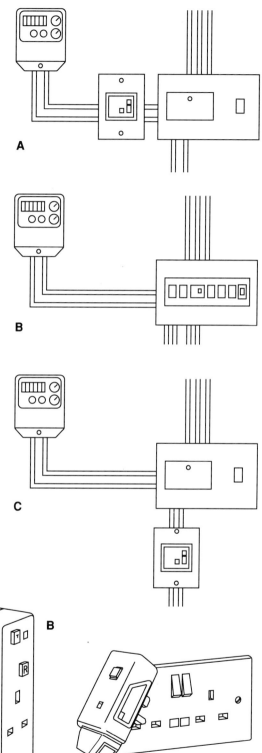

Fig 108 *Above*: The double-insulated symbol (one square inside another) means that a 2-core flex can be used for the appliance.

Fig 109 *Right*: Three ways of providing RCD protection to more than one socket outlet. **A** An RCD protecting the whole house; **B** A split-level consumer unit providing RCD protection to some circuits; **C** An RCD fitted into both wires of a socket outlet ring circuit.

Fig 110 *Below*: Three ways of providing RCD protection at a socket outlet. **A** An RCD-protected socket outlet (which replaces an ordinary socket outlet; **B** An RCD-protected adaptor which fits between plug and socket outlet; **B** An RCD-protected plug on the flex of equipment used outdoors.

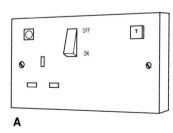
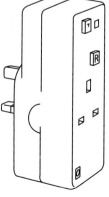
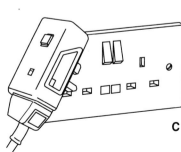

202 Safety in home electrics

(iii) Use the correct cable when extending the electricity supply into a detached garage, shed or greenhouse where the cable runs out of doors.

(iv) Do not use flex for fixed wiring.

(v) Do not have 'temporary' wiring hook-ups as these are not only potentially dangerous but tend to become permanent.

(vi) Do not up-rate fuses by using larger-size fuse wire.

(vii) Do not attempt to re-wire cartridge fuses or use substitute materials where no spare cartridge fuses are immediately available.

(viii) When wiring, double check that each lighting switch, which is single pole, is in the live pole of the circuit (see Fig 107). Reversal means that a ceiling rose terminal and a lampholder contact are still live when the switch is turned off.

(ix) Never use a radiant electric fire without the statutory guard which must be of small mesh to prevent a child inserting a finger.

(x) When using a mower or hedgecutters, or any power tool out of doors or in the garage or in similar situations, remember that you must use a high-sensitivity residual current device (RCD) fitted either in the socket outlet, in an adaptor or in the plug (see Fig 110) to ensure maximum personal protection from electrocution. This applies also to unearthed double-insulated tools, mowers and hedge-trimmers where, although double-insulated, exposed metal parts can become live if you cut the flex or drill into fixed wiring.

A better proposition is to have the RCD wired so that it protects the whole circuit, all socket outlet circuits or the whole house (see Fig 109).

(xi) Consider carefully before having an RCD installed in the lighting circuit, for a fault elsewhere could plunge the house into darkness and cause a serious or fatal accident such as falling down the stairs in the blackout.

(xii) In order to allow time to escape in the event of fire, fit battery-operated smoke alarms throughout the house preferably inter-linked so that when one is set off, all sound. Make sure at least one smoke alarm has an escape light.

Fig 111 *Top*: The swing label attached to electrical appliances which have been approved by the British Electrotechnical Approvals Board. *Below*: The equivalent mark on the appliance's rating plate.

Photo 114 A battery-operated smoke alarm is easily fitted to a ceiling to give advance warning in the event of fire (*Black & Decker*).

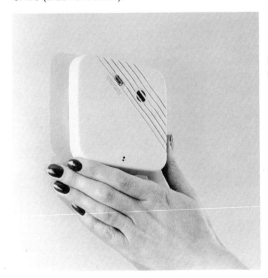

Index

Accidents, common causes, 197, 198
Appliances, electrical, *see* Fixed electrical appliances
Architrave switch, 80

Base unit heaters, 127
Bathroom safety, 199
Bathroom towel rail, installing, 130
Bell pushes, 137, 139
    illuminated, 137, 139
Bell transformers, installing, 138
Bells, electric, 137
    circuit wiring, 137
    extensions, circuits for, 139
    power sources for, 138
Bonding, *see* Earthing and bonding
Boxes, wiring accessory, 81, 85, 107
    architrave switch, 81, 82
    BESA used with light fittings, 44, 50, 52
    ceiling pattress, 44
    drylining, 86, 87
    dual flush for two socket outlets, 115
    dual surface for two socket outlets, 114
    flush 1-gang for socket outlets, 107
    flush, fitting to solid walls, 83, 84
    flush, fixing to dry partition wall, 85, 86
    lath and plaster partitions, fixing to, 86
    light switch, flush fixing, 82
    light switch, surface fixing, solid walls, 80
    mini-trunking, 34, 35, 36
    plateswitch, 1-gang, 81
    surface fixing, dry partition wall, 85
    surface 1-gang for socket outlets, 107
    surface 2-gang for double socket outlet, 113
Burglar alarms, 141
Buzzer, circuit wiring, 137
    power sources for, 138
Buzzers, electric, 137

Cable outlet unit, 143, 145, 146
Cables, housewiring, 29
    burying in wall plaster, 33, 84
    cutting and stripping, 30
    fishing under floorboards, 119
    fixing to wall surfaces, 33, 34
    fixings for, 33
    in mini-trunking, 34
    in roof space, 33
    installing, 30
    ratings for circuits (table), 37
    running inside partition wall, 34, 86
    single core, 29
    sizes, 30, 37
    types used in house, 30
    types used outside, 160, 161

    under floors, 30, 32
Ceiling roses, 38, 41, 42, 54, 55, 172
    loop-in, connecting flex to, 38
    loop-in, connection of extra cables, 55
    loop-in, three methods of wiring, 54
    replacing, 41
    wiring and fixing, 41, 55
Ceiling switches, 92
    45A version for shower unit, 149, 150
    double-pole, 92
    fixing and connecting, 91, 131
    light switch, 46
    replacing wall switch with, 91
    types, 92
Central heating, 128
Chimes, door, 138
    circuit wiring, 139
    portable, 140
    power sources for, 138
    transformers, installing, 138
Clock connectors, fused, 134, 135
Consumer units, 11, 178
    assembling after fixing to wall, 184
    choosing correct type and size, 180
    connecting circuit cables, 182
    fitted with MCB's, 180, 181
    fixing, 182
    for storage heaters, 157
    fuses for, 11, 12
    installing, 180, 182, 183
    isolating switch, circuit ratings of, 178
    materials required for installing, 180
    miniature circuit breakers, choice of, 14, 180
    RCD as the isolating main switch, 178
    RCD-protected, 17, 178
    removing old consumer unit, 182
    spare fuseways, provision for, 178
    split-load, 14, 178
    types of fuse, choice of, 178
Continuity tester, 13, 194
Contractors, choosing, 195
Cooker hood, fused connection unit for, 128
Cookers, electric, 142
    circuit cable sizes, 144
    circuit requirements, 142
    control switch without socket outlet, 142, 144, 146
    current rating of circuit, calculating, 142
    flush cooker control units, 145
    free-standing cooker, control unit and terminal box connections for, 143
    hob unit, built-in ceramic, 147
    installing a cable outlet unit, 145
    materials required for circuit wiring, 144
    planning and wiring the circuit, 144

204 Index

split-level cooker, wiring layout and cable connections at control switch, 143
surface type cooker control unit, 145
wiring and fixing control unit, 144
Cord-operated double-pole ceiling switch, cable connections, 131
Cornice trunking, 36, 37
Cylinders, hot water, *see* Hot water cylinders

Decorative lights, special fittings for, 43
Dimmer switches, 98, 99, 100
remote control, 100, 101
Direct hot water cylinders, 151
DIY wiring, regulations affecting, 194
Door chimes, *see* Chimes, door
Door entry telephone system, 140
Double-insulated symbol, 201
Double-pole (DP) switches, 92, 133, 145, 146, 148, 149, 168
Downlighters, 70
Drylining box, 86, 87
Dual switch, immersion heater control, 152

Earth-leakage circuit breakers, *see* Residual current devices
Earthing and bonding, 185
earth rod, 185, 186
equipotential bonding of gas and water services, 186, 187
metal lampholders, earthing, 40
protective multiple earthing (PME), 187
purpose, bonding, 186
purpose, earthing, 185
residual current device, 185, 186
supplementary bonding, 186, 188
wall lights, earthing, 50
warning label on earth clamp, 185
Economy 7 controller, 152, 153, 155
Economy 7 hot water cylinder, 151
Economy 7 tariff, 151, 191
Electricity company's requirements, 191
connection of new wiring, 194
DIY wiring, 194
obtaining a supply of electricity, 191
tariffs, *see* Electricity tariffs and consumption
testing an installation, 194
testing by electrical contractors, 194, 195
wiring regulations, importance of, 194
Electricity outside the house, *see* Outdoor electrics
Electricity tariffs and consumption, 191
consumption of electricity by various electrical appliances, 192, 194
economy in use of electriciy, 192
reading a meter, 191
Extractor fan, installing, 128, 129
Eyeball spotlights, 70, 72

Fixed electrical appliances, installing, 125, 175
base unit heater, 127
bathroom towel rail, 130, 133
central heating, 128
cooker hood, 128
cookers, *see* Cookers, electric
extractor fan, 128, 129
immersion heaters, *see* Immersion heaters
shaver socket outlet, 132
shaver supply units, 132, 133
shower pumps, 130
showers, *see* Shower units
storage heaters, *see* Storage heaters
under worktops, 136
undercupboard lighting, 130
wall-mounted heaters, 126
waste disposal units, 128
water heaters, 128, 131
water softener, 130
Flex outlet unit, 132, 133, 136
Flexible cords, 18
accidents, due to misuse of, 196, 197, 198
appliances fitting to, 19, 199
braided circular, 18
ceiling roses, connecting to, 38
circular PVC-sheathed, 18
connectors, 20, 21
curly, 18, 20
dangerous practices, using, 196
extending, 20, 21
extension reels, choosing, 20
faults occurring in, 196, 198
heat-resisting, 18, 152, 155
iron, fitting to, 19
kettle, fitting to, 19
lampholders, connecting to, 39
light fittings, renewing flex in, 38
loop-in ceiling roses, connecting to, 38, 41
parallel twin, 18
plugs, wiring, *see* Plugs
renewing flexes in light fittings, 38
safety measures, 196
sizes and applications, 22
twisted-twin, 196
types of, 18
unkinkable, 18
waterproof connector, 22
wiring, permanent, restrictions on, 199
Floorboards, 31
raising, 31
raising a board to fish cables beneath, 119
re-laying, preparations necessary before, 32
Fluorescent lighting, 72–76
circuits of different types, 73
circular tubes, 53, 72
colours of tubes, 76
compact lamps, 52, 73
diagnosis of faults, 76
how fluorescent lights work, 73
illuminated ceilings, tubes used in, 76
in workshops/garages, 163
maintenance, 76
miniature tubes, 72

Index

'quickstart' fittings and tubes, 73
replacing existing lights with, 47
replacing tubes, 76
straight tubes, 53, 72
straight (linear) fitting, interior of, 74
switchstart fittings and tubes, 73
undercupboard striplights, 75
Fused connection units, 27, 116, 122, 175
   circuits used for, 124
   connecting to cables, 125
   cooker hood supplied from, 128
   extractor fan supplied from, 128
   fixing to flush and surface mounting boxes, 124
   lighting supplied from, 136
   spur cable supplying ring circuit connections, 136
   switched version, 122
   undercupboard light supplied from, 130
   unswitched version, 124
   versions, various, 123
   waste-disposal unit supplied from, 128
   water heater supplied from, 128
   water softener supplied from, 130
   wiring a unit, 122
Fused spur box, *see* Fused connection units
Fuses, circuit, 11, 17, 178, 180, 183
   advantages, cartridge type, 12
   advantages, rewirable type, 12
   cartridge, 12, 178
   colour coding, 12, 17
   connection units, fuses in, 122
   consumer units, fuses in, 11, 178, 180, 183
   disadvantages, cartridge type, 12
   disadvantages, rewireable type, 12
   mending, 12
   plug, fuses in, 17, 23
   ratings and colour codes, 17
   rewireable, 12, 178
   testing cartridge fuses, 13
   why they blow, 16

Garage, light and power for, 163, 165
Garage and workshop, circuit wiring for, 165
Garden, RCD protecting socket outlets, 168, 169
   lighting, 169
   socket outlets, 167

Heaters, wall mounted, installing, 126
   storage, *see* Storage heaters
Hollow walls, fixings for, 85
Hot water cylinders, 151

Illuminated ceilings, 76
   ceiling depth necessary for, 76
   circuit wiring, 77
   layout with cables, 77
   lighting for, 76
   quantity of light, 76
   sections of, 77
Immersion heaters, 151
   choosing, 151
   connection of cables at terminals, 155
   control, 151
   double-pole wall-mounted switch, connections, 152
   dual change-over switch connections, 154
   Economy 7 controller, 152
   electrical circuits for, 152, 153
   fitting the heater, 152
   time switch for controlling an immersion heater, 152, 156
   wiring the circuit, 152
Indirect hot water cylinder, 151
Installation, electrical, inspecting, 170

Joists, drilling for cables, 30, 32
Junction boxes, 59, 60, 61, 120, 173
   connecting spur cables to 30A version, 120
   fixing to timber between joists, 59
   inserting 30A box into ring circuit cable, 120
   lighting, adding a cable, 61
   lighting circuit, junction box system, connections, 59
   lighting circuits, connections for 1 light and 1 switch, 59
   lighting circuits, 4-way box for, 60, 61, 173
   lighting circuits, 6-way box for, 61, 173
   method of wiring lighting circuits, 172
   ring circuits, used in, 118
   wiring lighting circuits using, 173

Kettle, electric, connecting flex to, 19
Kitchens, electric cookers in, 142

Lamps, electric, 52, 53
   fluorescent lamps, 52, 53, 72
   low-voltage, 52, 53
   tungsten halogen, 52, 53
Lampholders, 39, 40, 46, 47, 48, 173
   batten, 46, 47, 48, 173
   batten, fitting a, 46
   batten, types of, 46
   loop-in batten, wiring connections, 48
   metal, importance of earthing, 40
   non-pendant screwed type, 40
   pendant-type, connecting flex to, 39
   pre-wired pendant set, 41
   wiring, 39
Lath-and-plaster partitions, fixing accessories to, 86
Light fittings, 38
   bulkhead exterior light, 61
   close-mounted ceiling type, fixing, 45
   choosing your lamps, 52
   decorative light, installing, 42
   fixing close-mounted ceiling types, 45
   fluorescent, 72, 73, 74, 75, 76
   installing a BESA box for use with, 44
   plain pendant, wiring, 38
   porch light, 61
   using a special ceiling fitting, 44
   wall lights, *see* Wall lights

Lighting circuit extensions, diagrams of various methods, 58
    junction box method for extensions, 57, 58, 59
    loop-in method for extensions, 57, 58
Lighting circuit, maximum lights permitted, 54
Lighting circuit projects, 54
    adding extra light and switch, 54, 55
    back door light, 62, 64
    downlighters, 70
    eyeball spotlights, 72
    fluorescent lighting, 72; *see also* Fluorescent lighting
    garden spotlights, installing, 169
    illuminated ceilings, 76; *see also* Illuminated ceilings
    installing spotlights, 69; *see also* Spotlights
    loft lights, 77–79; *see also* Loft light and power in
    low-voltage lighting, 79
    outside light from ring circuit, 65
    porch lights, 62, 63
    recessed lights, 70
    2A lighting circuit, 175
    wiring outside lights, 62, 64
    wiring porch light and switch, 59
    wiring up a security light, 66
Lighting circuit wiring, 54, 172, 173
Lighting circuit wiring, cables used, 54
Lighting controls, 80
    2-way switch conversion for bedrooms, 95
    2-way switch used as a 1-way switch, 90
    3-way switching circuit, 96
    ceiling switches, 92
    converting 1-way switching to 2-way, 93
    dimmer switches, 99
    fitting flush-mounted light switch, 83, 84
    fitting surface-mounted light switch, 80
    hand-held remote controllers, 100, 101
    installing a twin (2-gang) lighting switch, 97
    new 2-way switching circuits, 95
    plateswitches, 80
    replacing a faulty 2-gang switch, 91
    replacing damaged rocker switches, 90
    replacing old switches with plateswitches, 87, 88
    replacing wall switch with ceiling switch, 91
    retractive ceiling switch, 92
    security light switch, 102
    time delay switches, 101
    timer switches, 103
Lighting, downlighters, *see* Downlighters
Lighting, fluorescent, *see* Fluorescent lighting
Lighting, low-voltage, *see* Low-voltage lighting
Lighting, wiring lighting circuits, 172
Loft, light and power in, 79
    circuit wiring for lighting, 78
    circuit wiring for power, 78
    running cables in, 33
    switches, cable connections at, 78
    switches, lighting on landing below, 78
    wiring layout for lighting and power, 78

Loop-in system of wiring lighting circuits, 172
Low-voltage lighting, 51, 52, 71, 79, 169
Luminaire supporting coupler, (LSC), 44, 50, 170

Mainswitch and fuse unit, installing, 180
Mainswitch gear, layout of, 179
Mainswitches and earthing, 178
Miniature circuit breakers (MCBS), 14, 15, 16, 178
Meter, reading a, 191
Mini-trunking, 34, 35, 36, 37, 83, 109, 119, 124

Night storage heaters, *see* Storage heaters

Outdoor electrics, 160
    cables, types used, 160, 161
    choosing socket outlet positions, 163
    connecting outdoor socket outlets to ring circuit, 166
    garage, light and power for, 163, 165
    garden lighting, 169
    garden socket outlets, 167
    garden spotlights, 168
    installing indoor sections of underground cable, 163
    installing outdoor socket outlets, 163
    light fittings outdoors, *see* Outside lights
    materials required, overhead cable run, 160
    materials required, underground cable run, 161
    outbuilding, light and power in, 163
    overhead cable, installing, 160, 162
    outdoor socket outlets, 163
    RCD garden socket outlets protected by, 168, 189
    RCD protection for circuits in outbuildings, 160
    spotlights in the garden, 169
    underground cable, installing, 161, 162
    wiring and fixing outdoor socket outlets, 166
    workshop, light and power in, 163, 165
Outdoor lights, 59, 64, 65, 67, 136
    backdoor outside light, 2-way switching, 64
    installing, 64
    porch light wiring, 59
    ring circuit, wired from, 66, 136
    security light, 64, 67

Plasterboard, mounting box for, 86
Plateswitches, 80
Plugs, 22, 23, 24, 25, 26, 27
    2-pin, 24
    'Easywire', 24
    fused, 23, 25, 27
    round-pin, non-fused, 23, 25, 104, 175
    types of, 23
    wiring, 22, 24, 25, 26, 27
Porch light, and switch, wiring for, 59
    fixing ceiling-mounted type, 63
    fixing wall-mounted type, 63
Protective multiple earthing (PME), 187

Radial power circuits, 177
    installing, 177

Recessed lights, 70, 72
Regulations affecting the home electrical wiring installation, 9, 194
   Electricity Supply Regulations, 9, 194
   IEE Wiring Regulations, 9, 195, 196
Remote control switching, 101
Replacing wall switch with a ceiling switch, 91
Residual current devices (RCDs), 17, 160, 168, 188, 189, 201
   adaptor type, 189, 201
   combined with MCB, 16, 184
   fitting, 190
   how connected to circuit wiring, 189, 201
   installing, 188
   plug type, 189, 201
   protecting circuits in outbuildings, 160
   protecting socket outlets, 170, 189
   types of, 17, 188
   what an RCD does, 17, 187
   whole-house, 188, 201
   with earth electrode, 185, 186
Retractive ceiling-mounted switch, 92
Rewiring a house, 170
   2A lighting circuit, 175
   cables in situ, inspecting, 171
   combined loop-in and junction box lighting circuits, 174
   consumer unit, *see* Consumer units
   domestic power circuits, 175
   earthing and bonding, *see* Earthing and bonding
   earthing arrangements, 172
   inspecting an installation, 171
   installing new consumer unit, 182
   junction box method of wiring lighting circuits, 173
   lighting circuits, wiring new, 172
   loop-in method of wiring lighting circuits, 172
   main switchgear, examining, 171
   number of circuits, 170
   planning rewiring, 170
   radial power circuits, installing, 177
   RCD protection, 170
   removing old consumer unit, 182
   ring circuits, installing, 175
   socket outlets, disposition of, 176
   socket outlets, recommended number, 170, 177
   switchfuse units, installing, 184
   wiring accessories, 171
Ring circuits, 175
   cable sizes for, 176
   connecting cable at 13A socket outlets, 106
   extending, 121
   installing, 175
   installing a flush double socket outlet, 108
   outside light, wired from, 65
   single storage electric heater supplied from, 159
   spurs, 116, 175
   spurs installing, 116, 136, 175; *see also* Spurs, ring circuit
   testing cables at a socket outlet before installing a spur cable, 116, 117
   wall lights wired from, 52
Rocker switches, various types, 89

Safety, home electrics, 196
   accidents, common causes of, 197, 198
   appliances, precautions when using, 199
   bathroom, special precautions necessary, 199
   BEAB approval label fixed to approved appliances, 202
   flexible cords, safety precautions, 196
   general precautions, 196
   importance of, 196
   pendant lampholder, 38
   RCD's use of as personal protection against electrocution, 201
   single-pole switch in live pole, importance of ensuring, 200
   wiring, precautions to take when installing, 199, 200, 202
Security lights, 64, 66, 67
Security light switches, 102
Service connector box, 185
Shaver socket outlet, for rooms, other than bathrooms, 132
Shaver supply units for bathrooms, 132
Shower pumps, 130
Shower units, 147
   circuit cable sizes, choosing, 148
   circuit wiring for, 148
   connecting and fixing double-pole cord-operated switch, 148, 149, 150
   cord-operated double-pole switches, 148–150
   materials required for, 148
   planning the circuit, 148
   wiring connections at, 150
   wiring the circuit, 149
Skirting trunking, 36, 37
Sleeving, green/yellow, 29, 185
Smoke detectors, 202
Socket outlets, 104, 105, 177
   connecting cables at 13A socket outlets, 106
   garden, 167
   installing a flush 13A socket outlet, 108
   metal finish, 105
   outdoor, 163
   recommended number, 170, 177
   shaver socket outlets, 132
   surface-mounted, 111
   switched double, 104, 105
   switched single, 104, 105
   terminal arrangement of a double 13A double socket outlet, 110
   terminal arrangement of a single 13A socket outlet, 109
   tester, 194, 195
   with neon, 105
Socket outlets, projects, 104
   adding a socket outlet or fused connection unit, 116

converting a flush-mounted single socket outlet to a flush-mounted double, 111, 112
converting a flush-mounted single socket outlet to a surface mounted double, 111, 113
converting a single socket outlet to a double, 109, 110, 111, 112
converting surface mounted socket outlets to flush versions, 107
converting a surface-mounted single socket outlet to a surface-mounted double, 109, 110
extending ring circuit, 121
exterior, 166, 167
garden socket outlets, installing, 167
installing outdoor socket outlets, 164
moving a socket outlet to a new position, 115
replacing damaged socket outlets, 104
replacing 1 single socket outlet by 2 singles, 112, 114, 115

Spotlights, 68
garden, 168, 169
installing, 69
outdoor spotlights, 168
track systems, 69
types of, 68

Spurs, ring circuit, 116, 175
checking whether spur can be connected to an existing socket outlet, 116, 117
connecting spur cable to ring circuit, 118, 119, 136
fused, 118, 136, 176
inserting joint box into ring circuit for spur connection, 120, 121
installing, 175
lighting supplied from, 136
looping spur cable from existing ring circuit socket outlet, 119
non-fused, 116, 118, 175

Storage heaters, 155, 156, 157
20A double-pole switch for, 159
25A twin double-pole switch for use with fan-assisted storage heater, 158
circuit wiring for, 157, 158
combined storage/convector heaters, 157, 159
fan-assisted storage heaters, 157
installing storage heaters, 157
single storage heater supplied from and connected to ring circuit, 159
sizes, choosing, 157

Switches, 57, 62, 80, 159
25A twin isolating switch for fan-assisted night storage heaters, 158
double-pole switch for fixed appliances, 133
double-pole 20A switch for controlling electric storage radiators, 159
dual change-over switch for immersion heater, connection of wires, 154
fixing a wall switch, 56
fixing surface switch to solid wall, 80
height for light switches, 80
lighting plateswitches, 80, 82
plateswitches, with screwless terminals, 82, 83
single-pole switch in live wire, 200
standard rocker plateswitches, 81
twin (2-gang) lighting switch, 97
wiring 2-way switch for outdoor light, 62, 65
wiring and fixing a light switch, 56

Switchfuse units, 184
connecting meter tails to, 184, 185
installing, 184

Switching, 95
2-way switch conversion for bedrooms, 95
intermediate switching circuit, 96
new 2-way circuits, 95
twin (2-gang) switching, 97
wiring 3-way (intermediate) switching circuit for lighting, 96

Tariffs, electricity, 191
Telephone system, door entry, 140
Testing, checking for adjacent socket outlets on ring circuit, 121
Testing, checking for spur cable connection at socket outlet, 116, 117
Testing, wiring, 9, 194
Time delay switch, 101
Timer switches, 103
for immersion heater, 152, 156, 192
Tools, types and selection of, 28, 29
Transformers, bell and chime, 138
Trunking for cables, *see* Mini-trunking

Undercupboard lighting, 130

Wall lights, 47
boxes for, 50, 52
fixing to wall, 50
independently switched, circuit wiring, 49
installing, 47, 50
not independently switched, circuit wiring, 49
planning the circuit wiring, 48
replacing an existing ceiling light, 50
special connectors for, 50, 51
wired into ring circuit, 52
wiring for, 47
Waste disposers, installing, 128
Water heaters, fused connection unit for, 128
Water softener, 130
Wiring, extra light and switch, 54
Wiring inspection and test, 194
regulations affecting DIY, 9, 10, 194
permission to install, 9
Workshop, lighting and power in, 163, 165